中国古建筑赏析

Appreciation of Chinese Traditional Architecture

主　编　刘嘉祎
副主编　石书羽　张　宁　刘　萍
参　编　艾嘉禾　董明明　王　丹
翻　译　钱　瑞

北京理工大学出版社
BEIJING INSTITUTE OF TECHNOLOGY PRESS

内容提要

本书为辽宁鲁班工坊项目教材（双语版），全面、系统地介绍了中国古建筑从结构与构造到表现形式的相关内容，涵盖了中国古建筑中的宫廷建筑、民居建筑、园林建筑、中南海建筑等知识点。本书共分5个模块、21个项目式教学单元。每个项目都融入了思政课程元素，使学生可以通过中国古建筑更深层次地了解中国传统文化。

本书既可作为职业院校学生通识教育类教材，也可作为相关行业社会人员及古建筑爱好者的参考用书。

版权专有　侵权必究

图书在版编目(CIP)数据

中国古建筑赏析：英文、汉文 / 刘嘉祎主编. --北京：北京理工大学出版社，2023.2
　ISBN 978-7-5763-1722-0

Ⅰ.①中… Ⅱ.①刘… Ⅲ.①古建筑－建筑艺术－中国－教材－汉、英 Ⅳ.①TU-092.2

中国版本图书馆CIP数据核字（2022）第173511号

出版发行 / 北京理工大学出版社有限责任公司	
社　　址 / 北京市海淀区中关村南大街5号	
邮　　编 / 100081	
电　　话 /（010）68914775（总编室）	
（010）82562903（教材售后服务热线）	
（010）68944723（其他图书服务热线）	
网　　址 / http://www.bitpress.com.cn	
经　　销 / 全国各地新华书店	
印　　刷 / 河北鑫彩博图印刷有限公司	
开　　本 / 787毫米×1092毫米　1/16	
印　　张 / 20.5	责任编辑 / 钟　博
字　　数 / 473千字	文案编辑 / 钟　博
版　　次 / 2023年2月第1版　2023年2月第1次印刷	责任校对 / 刘亚男
定　　价 / 108.00元	责任印制 / 王美丽

图书出现印装质量问题，请拨打售后服务热线，本社负责调换

前 言
PREFACE

本书是一本以中国古建筑为切入点,展现中国优秀历史传统的中英双语教材。本书立足"三教"改革,以职业教育发展新需求为抓手,在教材的理论体系、组织架构、内容描述方面都符合职业教育的教学需求、特点与规律。同时,本书充分体现了辽宁鲁班工坊的教育宗旨,并契合"一带一路"高等教育共同体的发展原则,推动"一带一路"沿线国家和地区大学之间在教育、文化等领域的全面交流与合作。

Taking Chinese traditional architecture as the entry point, *Appreciation of Chinese Traditional Architecture* is a Chinese-English bilingual textbook that aims at introducing excellent Chinese history and traditions. Based on the "Reform of teachers, textbooks and teaching methods", this textbook fully adheres to the new requirement of developing vocational education and fits the teaching requirements, features and laws of vocational education in terms of the theoretical system, organizational structure and content. In the meanwhile, this textbook fully reflects the educational tenets of Liaoning Luban Workshop, conforms the principle of developing a community of higher education along the Belt and Road and aims to advance the comprehensive communication and cooperation of universities of countries and regions along the Belt and Road in fields such as education and culture.

本书为项目式教材,共分为5个模块、21个项目式教学单元。第一部分为中国古建筑的结构与构造,以基础的结构与构造为切入点,介绍古建筑的人文价值和艺术价值,及其对现代建筑的影响与启发,使学生们充分理解鲁班精神与大国工匠精神之所在;第二部分为中国古建筑表现形式中的宫廷建筑,使学生们通过领略故宫(紫禁城)木质结构的恢宏大气,了解中国的皇家文化以及建筑内涵,感受中国传统文化的博大精深;第三部分为中国古建筑表现形式中的民居建筑,详细阐述了南方民居与北方民居的建筑制式与设计风格,使学生们通过民间建筑进一步感知古人的智慧与创造力;第四部分为中国古建筑表现形式中的园林建筑,使学生们通过园林建筑元素、组成要素以及皇家园林与私家园林,感受古人造园的智慧,领略园林艺术的美学;第五部分为中南海,叙述了古建筑的前世今生,并见证了中华人民共和国的成立。

As a project-based teaching material, this book can be divided into five chapters and twenty-one teaching units. The first chapter is about the structure and structure of Chinese traditional architecture. Starting from the basic structure and structure, it helps us understand the humanistic value and artistic value of ancient architecture, enlighten us about the creation of modern architecture, and fully understand the spirit of Luban and the workmanship spirit of our country. The second chapter introduces the palace architecture of Ancient Chinese architecture. By appreciating the magnificent wood architecture of the Forbidden City, readers can understand the royal culture and architectural connotation of China and experience the extensive and profound traditional Chinese culture. The third chapter introduces the residential architecture of Chinese ancient architecture, and elaborates the architectural system and design style of the folk architecture in the south and the north of China, which is conducive for learners to further perceive the wisdom and creativity of the ancients. The fourth chapter talks about the garden architecture of Ancient Chinese architecture. Through the garden architecture elements, components, as well as the royal garden and private garden, readers can feel the wisdom of ancient Chinese in building gardens and appreciate the aesthetics of garden art. Focusing on Zhongnanhai, the fifth chapter narrates the development of ancient buildings that can be seen as the witnesses of the founding of the People's Republic of China.

通过对中国古建筑的赏析，学生们可以认识到古建筑是人类遗产的实物表现、城市的凝固记忆，是城市意象的主体标志性符号。作为地域文化与灵魂的代表，可归属为文化遗产的古建筑，有着深厚的历史文化积淀和突出的建筑、艺术、科技以及美学上的价值。

Through the appreciation of Chinese ancient architecture, students can realize that ancient architecture is the expression of human heritage, the persistent memory of the city, and the main symbol of the city image. As the representative of regional culture and soul, ancient buildings can be classified as cultural heritage that with profound historical and cultural accumulation and prominent architectural, artistic, technological and aesthetic values.

编 者
Editor

目　录
CONTENTS

项目 1　从"墙倒屋不塌"说起

1.1　项目表单 ··· 001
1.2　理论指导 ··· 007
　　1.2.1　"墙倒屋不塌"的三个维度——建筑材料 ··· 007
　　1.2.2　"四梁八柱" ··· 007
　　1.2.3　榫卯结构 ·· 008
1.3　中华民族传统文化精神解读 ··· 008
1.4　项目创新 ··· 009

项目 2　台基与地面

2.1　项目表单 ··· 010
2.2　理论指导 ··· 017
　　2.2.1　古建筑基本结构 ··· 017
　　2.2.2　台基与台基的作用 ··· 017
　　2.2.3　台基的形式 ·· 018
　　2.2.4　踏跺及踏跺的表现形式 ·· 020
　　2.2.5　古建筑地面 ·· 022
2.3　中华民族传统文化精神解读 ··· 023
2.4　项目创新 ··· 023

项目 3　墙体构造

3.1　项目表单 ··· 024

3.2 理论指导 ······ 030
    3.2.1 古建筑墙体类型 ······ 030
    3.2.2 古建筑墙体形式 ······ 032
    3.2.3 古建筑墙体的作用 ······ 035
3.3 中华民族传统文化精神解读 ······ 035
3.4 项目创新 ······ 035

项目 4　木构架构造

4.1 项目表单 ······ 036
4.2 理论指导 ······ 042
    4.2.1 抬梁式木构架 ······ 042
    4.2.2 穿斗式木构架 ······ 042
    4.2.3 井干式木构架 ······ 043
4.3 中华民族传统文化精神解读 ······ 043
4.4 项目创新 ······ 043

项目 5　斗栱构造

5.1 项目表单 ······ 044
5.2 理论指导 ······ 049
    5.2.1 斗栱的含义 ······ 049
    5.2.2 斗栱的作用 ······ 049
5.3 中华民族传统文化精神解读 ······ 050
5.4 项目创新 ······ 050

项目 6　屋顶构造

6.1 项目表单 ······ 051
6.2 理论指导 ······ 057
    6.2.1 古建筑屋顶的形制 ······ 057
    6.2.2 重檐 ······ 058
    6.2.3 攒尖顶 ······ 058
    6.2.4 屋顶上的屋脊名称 ······ 058
    6.2.5 万字顶 ······ 059
6.3 中华民族传统文化精神解读 ······ 059
6.4 项目创新 ······ 059

项目 7　木装修

7.1　项目表单 ··· 060
7.2　理论指导 ··· 065
　　7.2.1　古建筑外檐装修 ··· 065
　　7.2.2　古建筑内檐装修 ··· 068
7.3　中华民族传统文化精神解读 ··· 070
7.4　项目创新 ··· 070

项目 8　彩画

8.1　项目表单 ··· 071
8.2　理论指导 ··· 076
　　8.2.1　彩画的等级 ··· 076
　　8.2.2　彩画的特点 ··· 076
　　8.2.3　彩画的作用 ··· 077
8.3　中华民族传统文化精神解读 ··· 077
8.4　项目创新 ··· 077

项目 9　宫廷建筑——紫禁城（故宫）

9.1　项目表单 ··· 078
9.2　理论指导 ··· 083
　　9.2.1　"样式雷"的概念 ··· 083
　　9.2.2　"样式雷"的起源 ··· 083
　　9.2.3　"样式雷"的分类 ··· 083
9.3　大国工匠精神解读 ··· 086
9.4　项目创新 ··· 086

项目 10　宫廷建筑——外朝三殿

10.1　项目表单 ··· 087
10.2　理论指导 ··· 093
　　10.2.1　故宫外朝三殿的名称 ··· 093
　　10.2.2　太和殿 ··· 093
　　10.2.3　中和殿 ··· 094
　　10.2.4　保和殿 ··· 094

    10.2.5 三大殿的"千龙吐水" 095
10.3 中华民族传统文化精神解读 095
10.4 项目创新 095

项目 11 宫廷建筑——屋脊十兽

11.1 项目表单 096
11.2 理论指导 101
    11.2.1 屋脊十兽概况 101
    11.2.2 屋脊十兽的名称 101
    11.2.3 屋脊十兽的寓意 101
11.3 中华民族传统文化精神解读 104
11.4 项目创新 104

项目 12 宫廷建筑——故宫的建筑色彩及象征

12.1 项目表单 105
12.2 理论指导 111
    12.2.1 故宫中的色彩初印象 111
    12.2.2 故宫中的色彩美学解析及象征意义 111
12.3 中华民族传统文化精神解读 115
12.4 项目创新 115

项目 13 小宅门——四合院

13.1 项目表单 116
13.2 理论指导 122
    13.2.1 四合院的分类 122
    13.2.2 宅门的制式 122
    13.2.3 影壁的制式 123
    13.2.4 四合院各组成部分的名称及功能 124
13.3 中华民族传统文化精神解读 126
13.4 项目创新 126

项目 14 大宅门——乔家大院

14.1 项目表单 127

14.2 理论指导 ·· 133
    14.2.1 乔家大院的基本建筑制式 ··· 133
    14.2.2 乔家大院的六院布局及建筑特色 ·· 133
14.3 中华民族传统文化精神解读 ·· 135
14.4 项目创新 ·· 135

项目 15　徽州民居

15.1 项目表单 ·· 136
15.2 理论指导 ·· 142
    15.2.1 徽州民居基本建筑形制 ·· 142
    15.2.2 徽州民居与四合院布局的区别 ·· 143
    15.2.3 徽州民居建筑的特点 ··· 143
15.3 中华民族传统文化精神解读 ·· 144
15.4 项目创新 ·· 145

项目 16　福建土楼

16.1 项目表单 ·· 146
16.2 理论指导 ·· 152
    16.2.1 福建土楼概览 ··· 152
    16.2.2 承启楼的建筑制式及建筑特色 ·· 152
    16.2.3 土楼的建筑材料及作用 ·· 152
    16.2.4 田螺坑土楼 ··· 153
    16.2.5 二宜楼 ··· 154
16.3 中华民族传统文化精神解读 ·· 155
16.4 项目创新 ·· 155

项目 17　园林中的建筑元素

17.1 项目表单 ·· 156
17.2 理论指导 ·· 162
    17.2.1 中国古典园林中的建筑元素及其特点 ··· 162
    17.2.2 中国四大名亭 ··· 163
    17.2.3 江南三大名楼 ··· 164
17.3 中华民族传统文化精神解读 ·· 165
17.4 项目创新 ·· 165

项目18　园林组成要素

18.1　项目表单 …………………………………………………………………… 166
18.2　理论指导 …………………………………………………………………… 172
18.3　中华民族传统文化精神解读 ……………………………………………… 175
18.4　项目创新 …………………………………………………………………… 175

项目19　皇家园林

19.1　项目表单 …………………………………………………………………… 176
19.2　理论指导 …………………………………………………………………… 182
　　19.2.1　皇家园林概览 ………………………………………………………… 182
　　19.2.2　颐和园的功能及主要建筑布局 ……………………………………… 182
　　19.2.3　颐和园中的主要建筑元素及功能和作用 …………………………… 182
19.3　中华民族传统文化精神解读 ……………………………………………… 185
19.4　项目创新 …………………………………………………………………… 186

项目20　私家园林

20.1　项目表单 …………………………………………………………………… 187
20.2　理论指导 …………………………………………………………………… 193
　　20.2.1　私家园林的建筑特点 ………………………………………………… 193
　　20.2.2　中国四大名园 ………………………………………………………… 193
　　20.2.3　园林中的构景手法 …………………………………………………… 193
　　20.2.4　园林中的植物 ………………………………………………………… 194
20.3　中华民族传统文化精神解读 ……………………………………………… 195
20.4　项目创新 …………………………………………………………………… 195

项目21　中南海

21.1　项目表单 …………………………………………………………………… 196
21.2　理论指导 …………………………………………………………………… 202
　　21.2.1　中南海建筑概览 ……………………………………………………… 202
　　21.2.2　中华人民共和国成立 ………………………………………………… 202

参考文献 ……………………………………………………………………………… 311

CONTENTS
目 录

Project 1 From "walls fall down, the house does not collapse"

1.1 Theoretical guidance 204
    1.1.1 The three dimensions of "walls fall down, the house does not collapse" – building materials 204
    1.1.2 "Four beams and eight columns" 205
    1.1.3 Mortise and tenon construction 206
1.2 Interpretation of the spirit of traditional Chinese culture 207
1.3 Project innovation 207

Project 2 Table base and ground

2.1 Theoretical guidance 208
    2.1.1 Basic structure of ancient buildings 208
    2.1.2 The role of table base and pedestal 209
    2.1.3 Form of pedestal 209
    2.1.4 Stomp and the expression of stomp 212
    2.1.5 Ancient building grounds 215
2.2 Interpreting the spirit of traditional Chinese culture 216
2.3 Project innovation 216

Project 3 Wall construction

3.1 Theoretical guidance ·········· 217
    3.1.1 Types of ancient architectural walls ·········· 217
    3.1.2 Classification of ancient building wall forms by material ·········· 220
    3.1.3 The role of the walls of ancient buildings ·········· 223
3.2 Interpretation of the spirit of traditional Chinese culture ·········· 224
3.3 Project innovation ·········· 224

Project 4 Wooden frame construction

4.1 Theoretical guidance ·········· 225
    4.1.1 Beam-bearing wooden frames ·········· 225
    4.1.2 Bucket-type wooden frames ·········· 226
    4.1.3 Well-dry wooden frames ·········· 227
4.2 Interpretation of the spirit of traditional Chinese culture ·········· 227
4.3 Project innovation ·········· 227

Project 5 Arch structure

5.1 Theoretical guidance ·········· 228
    5.1.1 The meaning of the arch ·········· 228
    5.1.2 The role of arch ·········· 228
5.2 Interpretation of the spirit of traditional Chinese culture ·········· 229
5.3 Project innovation ·········· 230

Project 6 Roof construction

6.1 Theoretical guidance ·········· 231
    6.1.1 Roof form of ancient buildings ·········· 231
    6.1.2 Overlapping eaves ·········· 232
    6.1.3 Saved pinnacle ·········· 232
    6.1.4 The names of the ridges on the roof ·········· 233
    6.1.5 Wan Zi Ding ·········· 233
6.2 Interpretation of the spirit of traditional Chinese culture ·········· 234
6.3 Project innovation ·········· 234

Project 7 Wood decoration

7.1 Theoretical guidance ··· 235
    7.1.1 Interior eaves decoration of ancient buildings ···················· 235
    7.1.2 Interior eaves decoration of ancient buildings ···················· 239
7.2 Interpretation of the spirit of traditional Chinese culture ············· 241
7.3 Project innovation ··· 241

Project 8 Color painting

8.1 Theoretical guidance ··· 242
    8.1.1 Grades of color painting ··· 242
    8.1.2 Characteristics of color painting ································ 242
    8.1.3 The role of color painting ··· 244
8.2 Interpretation of the spirit of traditional Chinese culture ············· 244
8.3 Project innovation ··· 244

Project 9 Palace Architecture — Forbidden City

9.1 Theoretical guidance ··· 245
    9.1.1 The Concept of "Yangshi Lei Archives" ····················· 245
    9.1.2 The Origin of "Yangshi Lei Archives" ······················· 245
    9.1.3 Classification of "Yangshi Lei Archives" ··················· 246
9.2 Interpretation of the spirit of traditional Chinese culture ············· 248
9.3 Project innovation ··· 249

Project 10 Palace architecture — the three great halls of the outer court

10.1 Theoretical guidance ·· 250
    10.1.1 The name of the "three great halls of the outer court" ········ 250
    10.1.2 Taihe Hall ··· 250
    10.1.3 Zhonghe hall ·· 252
    10.1.4 Baohe hall ·· 252
    10.1.5 The "Thousand Dragons Spitting Water" in the Three Great Halls ········ 252

| 10.2 | Interpretation of the spirit of traditional Chinese culture | 253 |
| 10.3 | Project innovation | 253 |

Project 11 Palace architecture — ten animal ornaments on house-ridges

11.1	Theoretical guidance	254
	11.1.1 The overview of "ten animal ornaments on house-ridges"	254
	11.1.2 The name of "ten animal ornaments on house-ridges"	254
	11.1.3 The meaning of "ten animal ornaments on house-ridges"	254
11.2	Interpretation of the spirit of traditional Chinese culture	258
11.3	Project innovation	258

Project 12 Palace architecture — the aesthetics and symbols of color in the architecture of the Forbidden City

12.1	Theoretical guidance	259
	12.1.1 First impression of color in the Forbidden City	259
	12.1.2 Aesthetic interpretation and symbolism of colour in the Forbidden City	260
12.2	Interpretation of the spirit of traditional Chinese culture	264
12.3	Project innovation	264

Project 13 Small House Gate — the quadrangle

13.1	Theoretical guidance	265
	13.1.1 The classifications of quadrangle	265
	13.1.2 The system of house gate	265
	13.1.3 Shadow wall system	267
	13.1.4 The name and function of each component of the quadrangle	268
13.2	Interpretation of the spirit of traditional Chinese culture	270
13.3	Project innovation	270

Project 14 Great House Gate — Qiao Family Compound

| 14.1 | Theoretical guidance | 271 |

    14.1.1 The basic architectural system of the Qiao Family Compound ·············· 271

    14.1.2 The six courtyard layout and architectural features of the Qiao
            Family compound ·· 271

14.2 Interpretation of the spirit of traditional Chinese culture ················ 275

14.3 Project innovation ·· 275

Project 15 Huizhou Folk Houses

15.1 Theoretical guidance ·· 276

    15.1.1 The basic architectural form of Huizhou houses ······························ 276

    15.1.2 The difference between Huizhou houses and courtyard layout ·········· 277

    15.1.3 The characteristics of Huizhou residential architecture ···················· 278

15.2 Interpretation of the spirit of traditional Chinese culture ················ 279

15.3 Project innovation ·· 280

Project 16 Fujian Tulou

16.1 Theoretical guidance ·· 281

    16.1.1 Overview of Fujian Tulou ·· 281

    16.1.2 Architectural style and architectural features of Chengqi building ·········· 281

    16.1.3 Building materials and functions of Tulou ·· 282

    16.1.4 Tianluokeng Tulou cluster ·· 282

    16.1.5 Eryi Building ·· 284

16.2 Interpretation of the spirit of traditional Chinese culture ················ 285

16.3 Project innovation ·· 285

Project 17 Architectural elements in the garden

17.1 Theoretical guidance ·· 286

    17.1.1 Architectural elements and characteristics in classical Chinese gardens ··· 286

    17.1.2 Four Famous Pavilions in China ·· 288

    17.1.3 Three Great Buildings of Southern China ·· 289

17.2 Interpretation of the spirit of traditional Chinese culture ················ 291

17.3 Project innovation ·· 291

Project 18   Elements of the garden

18.1 Theoretical guidance ········· 292
18.2 Interpretation of the spirit of traditional Chinese culture ········· 296
18.3 Project innovation ········· 296

Project 19   Royal Gardens

19.1 Theoretical guidance ········· 297
    19.1.1 Overview of the Royal Gardens ········· 297
    19.1.2 Functions and main architectural layout of the Summer Palace ········· 297
    19.1.3 The main architectural elements in the Summer Palace and their functional roles ········· 297
19.2 Interpretation of the spirit of traditional Chinese culture ········· 302
19.3 Project innovation ········· 302

Project 20   Private Garden

20.1 Theoretical guidance ········· 303
    20.1.1 Construction characteristics of private garden ········· 303
    20.1.2 The Four Famous Gardens of China ········· 303
    20.1.3 Scenic techniques in gardens ········· 304
    20.1.4 Plants in the garden ········· 305
20.2 Interpretation of the spirit of traditional Chinese culture ········· 306
20.3 Project innovation ········· 307

Project 21   Zhongnanhai

Theoretical guidance ········· 308
    1. Overview of Zhongnanhai Architectural ········· 308
    2. Establishment of New China ········· 309

References ········· 312

项目1 从"墙倒屋不塌"说起

1.1 项目表单

学习任务单

学习领域	"墙倒屋不塌"的原理				
学习情境	学习"墙倒屋不塌"的相关知识点				
学习任务	"墙倒屋不塌"的结构原理		学时	2学时(90分钟)	
教学目标	使学生充分了解"墙倒屋不塌"的原理				
技能目标	通过"墙倒屋不塌"进一步了解中国古建筑结构的精髓及古人的智慧				
知识目标	掌握建筑材料、"四梁八柱"及榫卯结构的相关知识点				
学习目标	1. 了解建筑材料的特点; 2. 了解"四梁八柱"的含义; 3. 了解斗栱的作用; 4. 了解开间的含义; 5. 了解榫卯结构的含义与作用				
任务描述	从3个维度熟练掌握"墙倒屋不塌"的原理				
学时安排	情境导入	计划	决策	实施	检查	评价
时长	10分钟	10分钟	10分钟	30分钟	15分钟	15分钟
对学生的要求	1. 课前做好项目预习; 2. 做好课中总结与评价; 3. 通过"墙倒屋不塌"的学习内容谈"建筑语言中的人文主义精神"				
参考资料	1. 微课视频; 2. PPT; 3. 课程相关视频资料				

项目前学习任务单

学习领域	"墙倒屋不塌"的原理	用时	
达成目标	1. 观看微课视频，可以简单整理出课程内容思维导图； 2. 根据课程内容，感受中国古建筑文化的博大精深及大国工匠精神		
学习方法	看：观看微课视频及相关课程资料； 思：认真思考任务单中的问题； 记：将心中的疑惑记下来		
课堂学习形式预告	课堂外 1. 自主学习，搜集资料； 2. 整理收获，提出问题		课堂内 1. 展示交流； 2. 协作探究； 3. 启发式教学； 4. 项目驱动； 5. 实习作业法
本次学习任务	任务1：了解建筑材料的特点； 任务2：了解"四梁八柱"的含义； 任务3：了解斗栱的作用； 任务4：了解开间的含义； 任务5：了解榫卯结构的含义与作用		
疑点与问题	本次学习中还有哪些不明白的地方？请依次记录下来。 		
总结评价	1. 你认为自己预习得怎么样？（　　） 　A. 马马虎虎，只是为了完成任务 　B. 态度认真，收获很多 2. 你对自己本次自主学习的评价：☆☆☆☆☆		

资讯单

学习领域	"墙倒屋不塌"的原理		
学习情境	学习"墙倒屋不塌"的相关知识点		
学习任务	"墙倒屋不塌"的结构原理	学时	10分钟
资讯方式	1. 教师讲解； 2. 互联网查询； 3. 同学交流		
资讯问题	1. 古建筑的建筑材料是什么？ 2. 古建筑的建筑材料有哪些特点？ 3. "四梁八柱"的含义是什么？ 4. 斗栱的作用有哪些？ 5. 什么是开间？ 6. 通过"墙倒屋不塌"的学习内容谈"大国工匠精神"		
资讯引导	1. 微课视频； 2. PPT； 3. 课程相关视频资料		

计划单

学习领域	"墙倒屋不塌"的原理		
学习情境	学习"墙倒屋不塌"的相关知识点		
学习任务	"墙倒屋不塌"的结构原理	学时	10分钟
典型工作过程描述	教学情境导入—建筑材料—"四梁八柱"—斗栱—开间—榫卯结构—民族文化精神解读		
计划制订的方式	同学间分组讨论		
序号	工作步骤	注意事项	
1	查看资料文件		
2	查询资料		
3	设计课程内容思维导图		
计划评价	班级	第___组	组长签字
	教师签字	日期	
	评语：		

决策单

学习领域	"墙倒屋不塌"的原理			
学习情境	学习"墙倒屋不塌"的相关知识点			
学习任务	"墙倒屋不塌"的结构原理	学时		10分钟
典型工作过程描述	教学情境导入—建筑材料—"四梁八柱"—斗栱—开间—榫卯结构—民族文化精神解读			
学生为主体的课堂决策				
序号	课程目标	课程内容	教学活动	教学资料	综合评价
1					
2					
3					
4					
5					
6					
7					
8					
9					
10					
	班级		第___组	组长签字	
	教师签字		日期		
计划评价	评语:				

实施单

学习领域	"墙倒屋不塌"的原理		
学习情境	学习"墙倒屋不塌"的相关知识点		
学习任务	"墙倒屋不塌"的结构原理	学时	30分钟
典型工作过程描述	教学情境导入—建筑材料—"四梁八柱"—斗栱—开间—榫卯结构—民族文化精神解读		

序号	实施步骤	注意事项
1	教学情境导入	播放先导片
2	古建筑材料的特点	熟练掌握
3	"四梁八柱"的含义	教学重点与难点
4	斗栱的作用	熟练掌握
5	开间的含义	熟练掌握
6	结合课程思政	引入"中华民族传统文化精神"

实施说明：
1. 播放先导片；
2. 借助微课视频与PPT进行深入了解；
3. 通过了解"墙倒屋不塌"的原理引入"中华民族传统文化精神"

实施评价	班级		第___组	组长签字	
	教师签字		日期		
	评语：				

检查单

学习领域	"墙倒屋不塌"的原理			
学习情境	学习"墙倒屋不塌"的相关知识点			
学习任务	"墙倒屋不塌"的结构原理		学时	15分钟
典型工作过程描述	教学情境导入—建筑材料—"四梁八柱"—斗栱—开间—榫卯结构—民族文化精神解读			
序号	检查项目	检查标准	学生自查	教师检查
1	资讯环节	获取相关信息情况		
2	计划环节	设计课程思维导图		
3	实施环节	绘制课程思维导图		
4	检查环节	各个环节逐一检查		
检查评价	班级		第___组	组长签字
	教师签字		日期	
	评语：			

评价单

学习领域	"墙倒屋不塌"的原理			
学习情境	学习"墙倒屋不塌"的相关知识点			
学习任务	"墙倒屋不塌"的结构原理		学时	15分钟
典型工作过程描述	教学情境导入—建筑材料—"四梁八柱"—斗栱—开间—榫卯结构—民族文化精神解读			
评价项目	评价子项目	学生自评	组内评价	教师评价
资讯环节	1. 听取教师讲解； 2. 互联网查询情况； 3. 同学交流情况			
计划环节	1. 查询资料情况； 2. 设计思维导图			
实施环节	1. 学习态度； 2. 课程内容熟练程度； 3. 课堂表现； 4. 作业完成情况及质量			
最终结果	综合情况			
评价	班级		第___组	组长签字
	教师签字		日期	
	评语：			

1.2 理论指导

1.2.1 "墙倒屋不塌"的三个维度——建筑材料

从"墙倒屋不塌"说起

"墙倒屋不塌"是对中国传统古建筑最形象的描述，其中的奥秘就隐藏在建筑材料和结构中。可以从建筑物的三个维度来揭秘"墙倒屋不塌"的原理，这三个维度分别是建筑材料（木材）、"四梁八柱"及榫卯结构。

木材（图1-1）的特点如下。

（1）木材具有质量小、柔韧性强的特点，同时木结构自重较轻。

（2）木构件便于运输、装拆，可多次使用。

（3）木材还具有防火及防腐性，因此常被用作建造房屋的材料。

图1-1 木材

1.2.2 "四梁八柱"

中国古建筑无论规模如何，其外观轮廓均由台基、梁柱木造部分及屋顶组成（图1-2）。古代多数建筑采用三开间的格局，由于开间较大，要在中间加两道梁，加上前梁、后梁，合计是4根梁，每根梁的两端各有一根柱子，起到支撑的作用。这样就有8根柱子。靠这4根梁和8根柱子支撑整个建筑。其中墙面是用来分隔空间的，不是主要起到支撑作用。这样就有了"四梁八柱"之说，代表主要结构（图1-3）。

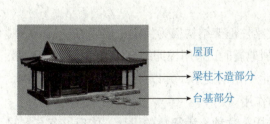

图1-2 中国古建筑结构

图1-3 中国古建筑内部结构（"四梁八柱"）

1. 斗栱

斗栱作为木结构部件，它的作用是伸出悬臂承托出檐部分的重量（图1-4）。

2. 开间

开间就是指正面相邻两檐柱间的距离，也称"面阔"。檐柱就是上面4个图形覆盖的区域，也就是房屋最外面的柱子（图1-5）。中国古建筑中的开间一般都是奇数开间，如一开间、三开间、五开间、七开间、九开间。

拥有开间最多的中国古建筑是故宫中的太和殿，它有 11 间开间，是中国古建筑中的最高建筑制式。民间建筑一般都用三开间或五开间，因为民间建筑规模较小。

图 1-4 斗栱

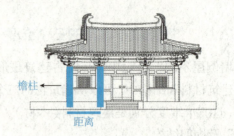

图 1-5 古建筑开间与檐柱示意

1.2.3 榫卯结构

榫卯结构在中国古建筑中是灵魂一般的存在（图 1-6）。榫卯是中国古代木构件利用凹凸相互扣合实现连接的一种独特方式。"凸榫凹卯"，在榫卯结构中，榫是木构件凸出的部分，又叫作榫头。榫头伸入卯眼的部分称为榫舌，其余部分则称为榫肩。卯是凹入的部分，又称为榫眼、榫槽。榫和卯相互咬合，形成产生"1+1>2"的支撑力量。

榫卯被广泛应用于建筑、家具及其他木制器械。榫卯技艺在明清时达到最高水平。

榫卯结构在建造史上无法超越，不用一钉一胶，在木头上分别修出彼此契合的形状，两者即可以结合在一起，相互咬合，以达到稳固的作用。

制作榫卯结构也是古代木匠必须具备的基本技能，工匠手艺的高低，通过榫卯结构就能清楚地反映出来。被尊为中国木匠鼻祖的鲁班将榫卯结构发挥到极致，榫与卯间渗透着大国工匠精神。

图 1-6 榫卯结构

榫卯结合是抗震的关键，这种结构和现代梁柱框架结构极为类似，构架的节点不是刚性的，这就保证了建筑物的刚度协调。遇有强烈地震时，采用榫卯结合的空间结构虽会"松动"，却不致"散架"，它允许房屋的结构产生一定的变形，并以此消耗地震传来的能量，使整个房屋的地震荷载大为减小，起到抗震的作用。

中国古代匠人创造出精巧的斗栱及榫卯结构，这种充满智慧的设计，体现了东方人的哲学思维方式——以柔克刚。

1.3 中华民族传统文化精神解读

中国古建筑的"墙倒屋不塌"可谓建筑史上的一个奇迹，无论是古人对木质材料的钟爱还是"四梁八柱"支撑起整个建筑，再到精密的榫卯结构，每一处都散发着古人的智慧，蕴含了中华文化的博大精深，那些结构精巧的建筑诉说着来自古老文明的鲜明的人文主义精神。

1.4 项目创新

　　结合课程内容中的"墙倒屋不塌"原理，谈一谈中国古建筑的精髓所在，以及对古人建造智慧的理解与感想。

项目 2　台基与地面

2.1　项目表单

学习任务单

学习领域	台基与地面		
学习情境	学习台基与地面的相关知识点		
学习任务	古建筑台基及台基的作用、表现形式、踏跺及踏跺的表现形式	学时	2学时（90分钟）
教学目标	使学生充分了解古建筑台基与地面的相关知识点		
技能目标	通过古建台基与地面进一步了解中国古建筑结构的精髓及古人的智慧		
知识目标	掌握古建筑台基及台基的作用、表现形式，踏跺及踏跺的表现形式的相关知识点		
学习目标	1. 熟练掌握古建筑结构组成部分； 2. 熟练掌握古建筑台基的形式与作用； 3. 熟练掌握踏跺的表现形式； 4. 熟练掌握古建筑地面的分类		
任务描述	熟练掌握古建筑台基与地面的相关知识点		

学时安排	情境导入	计划	决策	实施	检查	评价
时长	10分钟	10分钟	10分钟	30分钟	15分钟	15分钟

对学生的要求	1. 课前做好项目预习； 2. 做好课中总结与评价； 3. 通过古建筑台基与地面的学习内容谈建筑语言中的人文主义精神
参考资料	1. 微课视频； 2. PPT； 3. 课程相关视频资料

项目前学习任务单

学习领域	台基与地面		用时	
达成目标	1. 观看微课视频，可以简单整理出课程内容思维导图； 2. 根据课程内容，感受中国古建筑文化的博大精深以及大国工匠精神			
学习方法	看：观看微课视频及相关课程资料； 思：认真思考任务单中的问题； 记：将心中的疑惑记下来			
课堂学习形式预告	课堂外		课堂内	
	1. 自主学习，搜集资料； 2. 整理收获，提出问题		1. 展示交流； 2. 协作探究； 3. 启发式教学； 4. 项目驱动； 5. 实习作业法	
本次学习任务	任务1：了解古建筑结构； 任务2：熟练掌握古建筑台基的形式与作用； 任务3：熟练掌握踏跺的表现形式； 任务4：熟练掌握古建筑地面的分类			
疑点与问题	本次学习中还有哪些不明白的地方？请依次记录下来。			

总结评价	1. 你认为自己预习得怎么样？
	  A. 马马虎虎，只是为了完成任务
	  B. 态度认真，收获很多
	2. 你对自己本次自主学习的评价：☆☆☆☆☆			

资讯单

学习领域	台基与地面		
学习情境	学习台基与地面的相关知识点		
学习任务	台基的概念形式与作用，踏跺的概念和表现形式，古建筑地面	学时	10分钟
资讯方式	1. 教师讲解； 2. 互联网查询； 3. 同学交流		
资讯问题	1. 古建筑的结构有哪几部分？ 2. 什么是台基？ 3. 台基的表现形式有哪些？ 4. 台基的作用是什么？ 5. 什么是踏跺？踏跺的表现形式有哪些？ 6. 通过台基与地面的学习内容谈中国古建筑中暗藏的民族智慧		
资讯引导	1. 微课视频； 2. PPT； 3. 课程相关视频资料		

计划单

学习领域	台基与地面		
学习情境	学习台基与地面的相关知识点		
学习任务	台基的概念形式与作用，踏跺的概念和表现形式，古建筑地面	学时	10分钟
典型工作过程描述	教学情境导入—古建筑结构—台基与台基的形式—台基的作用—踏跺与踏跺的表现形式—古建筑地面—民族文化精神解读		
计划制订的方式	同学间分组讨论		

序号	工作步骤	注意事项
1	查看资料文件	
2	查询资料	
3	设计课程内容思维导图	

计划评价	班级		第___组	组长签字	
	教师签字		日期		
	评语：				

项目 2 台基与地面

决策单

学习领域	台基与地面		
学习情境	学习台基与地面的相关知识点		
学习任务	台基的概念形式与作用，踏跺的概念和表现形式，古建筑地面	学时	10 分钟
典型工作过程描述	教学情境导入—古建筑结构—台基与台基的形式—台基的作用—踏跺与踏跺的表现形式—古建筑地面—民族文化精神解读		

学生为主体的课堂决策					
序号	课程目标	课程内容	教学活动	教学资料	综合评价
1					
2					
3					
4					
5					
6					
7					
8					
9					
10					

决策评价	班级		第___组	组长签字	
	教师签字		日期		
	评语：				

实施单

学习领域	台基与地面		
学习情境	学习台基与地面的相关知识点		
学习任务	台基的概念形式与作用，踏跺的概念和表现形式，古建筑地面	学时	30 分钟
典型工作过程描述	教学情境导入—古建筑结构—台基与台基的形式—台基的作用—踏跺与踏跺的表现形式—古建筑地面—民族文化精神解读		

序号	实施步骤	注意事项
1	教学情境导入	播放先导片
2	古建筑结构	熟练掌握
3	台基的形式与作用	教学重点与难点
4	踏跺的表现形式	教学重点与难点
5	古建筑地面	教学重点与难点
6	结合课程思政	引入中华民族传统文化精神

实施说明：
1. 播放先导片；
2. 借助微课视频与 PPT 进行深入了解；
3. 通过了解古建筑台基与地面引入中华民族传统文化精神

实施评价	班级		第___组	组长签字	
	教师签字		日期		
	评语：				

检查单

学习领域	台基与地面		
学习情境	学习台基与地面的相关知识点		
学习任务	台基的概念形式与作用，踏跺的概念和表现形式，古建筑地面	学时	15分钟
典型工作过程描述	教学情境导入—古建筑结构—台基与台基的形式—台基的作用—踏跺与踏跺的表现形式—古建筑地面—民族文化精神解读		

序号	检查项目	检查标准	学生自查	教师检查
1	资讯环节	获取相关信息情况		
2	计划环节	设计课程思维导图		
3	实施环节	绘制课程思维导图		
4	检查环节	各个环节逐一检查		

检查评价	班级		第___组	组长签字	
	教师签字		日期		
	评语：				

评价单

学习领域	台基与地面			
学习情境	学习台基与地面的相关知识点			
学习任务	台基的概念形式与作用，踏跺的概念和表现形式，古建筑地面		学时	15 分钟
典型工作过程描述	教学情境导入—古建筑结构—台基与台基的形式—台基的作用—踏跺与踏跺的表现形式—古建筑地面—民族文化精神解读			
评价项目	评价子项目	学生自评	组内评价	教师评价
资讯环节	1. 听取教师讲解； 2. 互联网查询情况； 3. 同学交流情况			
计划环节	1. 查询资料情况； 2. 设计思维导图			
实施环节	1. 学习态度； 2. 课程内容熟练程度； 3. 课堂表现； 4. 作业完成情况及质量			
最终结果	综合情况			
评价	班级		第___组	组长签字	
	教师签字		日期		
	评语：				

2.2 理论指导

2.2.1 古建筑基本结构

台基与地面

从造型上看一栋古建筑明显分为台基、屋身、屋顶3个部分。

北宋著名匠师喻皓在《木经》中称之为"三分",并指出"凡屋有三分,自梁以上为上分,地以上为中分,阶为下分"。

根据古建筑房屋这三部分的构成机能,古建筑构造可以分为以下几个部分。上分——屋顶;中分——屋身木构架、斗栱、墙体;下分——基础、台基、地面(图2-1)。

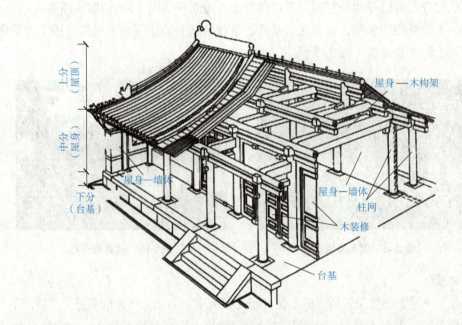

图2-1 古建筑三部分的构成机能

2.2.2 台基与台基的作用

台基是高出地面的建筑物底座,是用砖石砌成的突出的平台,用以承托建筑物,并使其防潮、防腐,同时可弥补中国古建筑单体建筑不甚高大雄伟的欠缺,最早是为了防潮、防水,后来则为了满足外观及等级制度的需要(图2-2)。

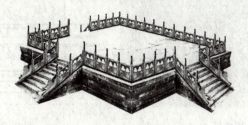

图2-2 台基

2.2.3 台基的形式

台基依据其使用功能和外形，大体分为普通台基和须弥座两类。

1. 普通台基

从早期遗迹来看，最早的台基全部由夯土筑成，推测至少在商代就已经出现（图2-3）。由夯土制成的台基又叫作夯土台。夯土就是将泥土压实，减少泥土之间的缝隙，增加泥土混合物的密度。

图2-3 商代夯土台

从汉代起夯土台已开始外包砖石来代替夯土，因此叫作磉（sǎng）墩（图2-4）。

还有一种较高级的台基，较普通台基高，常在台基上边建汉白玉栏杆，用于大型建筑或宫殿建筑中的次要建筑（图2-5）。

图2-4 汉代磉墩

图2-5 较高级台基

2. 须弥座

"须弥"一词原是梵文音译，相传是古印度神话中的名山。须弥的意思是"妙高""妙光""善积"等。因此，须弥山有时又被译为"妙高山"，意思是宝山。须弥座又名"金刚座"或"须弥坛"，后来代指建筑装饰的底座。

在封建社会中，须弥座只能使用在高等级的宫殿、寺院、道观等，以及一些纪念性建筑上。因此，须弥座被打上了深深的等级烙印（图2-6～图2-9）。

图2-6 须弥座

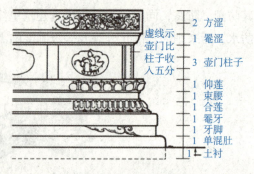

图2-7 宋式须弥座

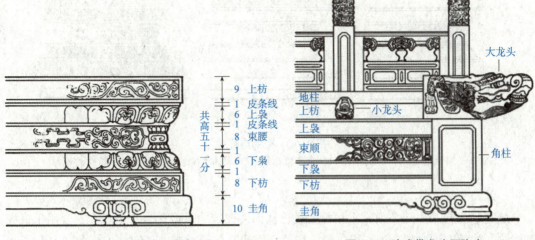

图 2-8 清式须弥座　　　　图 2-9 清式带龙头须弥座

故宫太和殿中的龙头须弥座里的龙头是一个排水系统，每逢大雨时节会展现出"千龙吐水"的奇观（图 2-10）。

图 2-10 故宫中"千龙吐水"奇观

须弥座的结构及样式如下。

清式须弥座一般为石制，自上而下分别为上枋（fāng）、上枭（xiāo）、束腰、下枭、下枋及圭角，它的雕刻装饰以束腰为基准也可分为单纯束腰部位的雕刻、束腰及上坊部位的雕刻以及全部雕刻的须弥座（图 2-11、图 2-12）。

图 2-11 须弥座结构

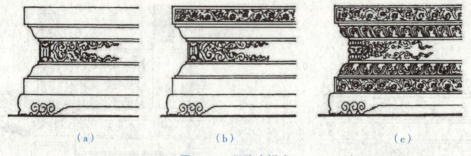

图 2-12 须弥座样式

(a) 束腰部位雕刻的须弥座；(b) 束腰和上枋部位雕刻的须弥座；(c) 全部雕刻的须弥座

3. 最高等级台基

最高等级台基由几个须弥座相叠而成，从而使建筑物显得更为宏伟高大，常用于最高级建筑，如故宫太和殿和天坛祈年殿，即耸立在最高等级台基上（图 2-13、图 2-14）。

图 2-13 故宫太和殿中的最高等级台基

图 2-14 天坛祈年殿中的最高等级台基

台基在中国古建筑中不仅有悠久的历史，而且范围十分广泛，上自宫殿，下至民宅，都可以见到它的存在。这种现象自然与使用台基的种种好处有关。

台基在结构上有承重作用，可以更好地承担上部的负荷，防止不均匀沉降的发生。由于台基的建造在生产力低下的时代是一件工程量很大和技术要求高的工作，所以有无台基和台基的高矮很自然地成为人们身份、地位的标志。

4. 丹陛石

丹陛石又称为陛阶石，是古代宫殿门前台阶中间镶嵌的那块长方形大石头，一般是一整块石头，常人不能在丹陛石上走，只能从它的两侧走，因为它是皇权的象征。

5. 丹陛石浮雕图案的象征意义

丹陛石上的浮雕图案，上部是宝珠和双龙，下部有海水和山石，其中宝珠是皇权的象征，两龙中，一龙象征天帝，另一龙象征皇帝，意谓皇权神授；海水和山石纹是江山永固的象征（图 2-15）。

图 2-15 丹陛石浮雕图案象征意义示意

2.2.4 踏跺及踏跺的表现形式

踏跺是中国古建筑中的台阶，一般用砖或石条砌造，置于台基与室外地面之间。踏跺的样式包括垂带踏跺、如意踏跺、礓磋（jiāng cā）石及御路踏跺（图2-16）。

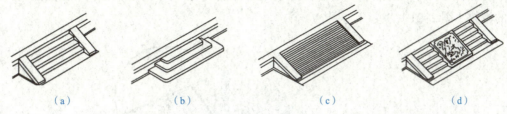

图 2-16　踏跺样式

（a）垂带踏跺；（b）如意踏跺；（c）礓磋石；（d）御路踏跺

1. 垂带踏跺

垂带踏跺是一种很工整的台阶，两边各有一条斜放的条石，叫作"垂带石"（图2-17）。

还有一种更高级的垂带踏跺叫作三出陛（图2-18）。在古代，高台式建筑是重要的皇家建筑，为了便于上下高台，也要修筑台阶，这种高台所用石台阶就是"陛"。为了分尊卑，上下高台的石台阶要有三路，按左、中、右排列，叫作"三出陛"。正中之陛，较宽，路中有带雕刻的丹陛石，两侧设有窄台阶；左、右两路陛等宽，但都比中路陛窄一些，供官员行走。

图 2-17　垂带踏跺

图 2-18　三出陛

2. 如意踏跺

如意踏跺是每层石板的长度和宽度都逐层缩小的台阶，人们可以从不同的方向上下，各面都设有其他维护设施（图2-19、图2-20）。

图 2-19　如意踏跺

图 2-20　天然如意踏跺

3. 礓磜石

礓磜石是一种不砌台阶而用石板或砖头铺成的斜面。为了防滑,在斜石面上凿刻出水平的模纹(图2-21)。

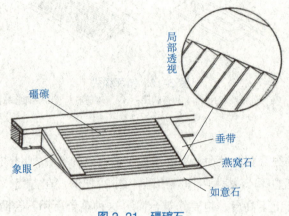

图2-21 礓磜石

4. 御路踏跺

御路踏跺一般用于宫殿与寺庙建筑,原为中国宫殿建筑形制,是位于宫殿中轴线上台基与地坪以及两侧阶梯间的坡道;在封建时代只有皇帝才能使用。它的中间有丹陛石,两侧有栏杆(图2-22)。

5. 龙尾道

龙尾道起源于战国时期的高台建筑,随着建筑高度的降低,其道前高后卑,下榻于地,宛如龙尾下垂,因此称为龙尾道(图2-23)。

图2-22 御路踏跺

图2-23 龙尾道

2.2.5 古建筑地面

古建筑地面包括甬路及散水。

1. 甬路

甬路是指用砖石砌成的路(图2-24)。

甬路根据地面砖石的细腻程度可分为糙墁地面及细墁地面。这里"墁"是涂抹、粉饰的意思。

图2-24 甬路

糙墁地面是指砖料不须砍磨加工，地面砖的接缝较宽，砖与砖相邻处的高低差和地面的平整度都不如细墁地面那样考究（图2-25）。

细墁地面是指砖料应经过砍磨加工，加工后的砖规格统一准确、棱角完整挺直、表面平整光洁。其中最精细的是"金砖墁地"，一般只用在宫殿等最重要的建筑（图2-26）。

图2-25　糙墁地面

图2-26　细墁地面

2. 散水

散水是指房屋外墙四周，室外地坪上，有一定坡度的散水坡。散水的作用是迅速排走房屋附近的雨水，避免雨水冲刷或渗透到地基，防止基础下沉，以保证房屋巩固耐久（图2-27）。

虎头找　宝剑头　燕尾　大岔　条砖芽子

图2-27　中国古建筑散水

2.3　中华民族传统文化精神解读

通过对古建筑台基与地面的了解，我们深刻感受到古代劳动人民的智慧。悠久的历史创造了灿烂的中华文化，而古建筑是其重要组成部分。在泥土铸成的台基与脚下的踏跺间仍然没有离开平面铺展的理性精神的基本线索，依然是把空间意识转化为时间过程，渲染表达的仍然是现实世间的生活意绪。

2.4　项目创新

通过台基与地面的学习内容，写出对中国古建筑中暗藏的民族智慧的感想，形式不限。

项目 3　墙体构造

3.1　项目表单

学习任务单

学习领域	墙体构造				
学习情境	学习墙体构造的相关知识点				
学习任务	古建筑墙体类型、墀头与墀头的作用、按照材料划分的墙体种类		学时	2学时（90分钟）	
教学目标	使学生充分了解古建筑墙体类型及按照材料划分的墙体种类				
技能目标	通过墙体构造进一步了解中国古建筑结构的精髓及古人的智慧				
知识目标	掌握古建筑墙体类型、墀头与墀头的作用、按照材料划分的墙体种类的相关知识点				
学习目标	1. 熟练掌握古建筑墙体类型； 2. 熟练掌握墀头与墀头的作用； 3. 熟练掌握按照材料划分的墙体种类				
任务描述	熟练掌握古建筑墙体构造的相关知识点				
学时安排	情境导入	计划	决策	实施	检查	评价
时长	10分钟	10分钟	10分钟	30分钟	15分钟	15分钟
对学生的要求	1. 课前做好项目预习； 2. 做好课中总结与评价； 3. 通过古建筑墙体构造的学习内容谈建筑语言中的人文主义精神				
参考资料	1. 微课视频； 2. PPT； 3. 课程相关视频资料				

项目前学习任务单

学习领域	墙体构造		用时	
达成目标	1. 观看微课视频，可以简单整理出课程内容思维导图； 2. 根据课程内容，感受中国古建筑文化的博大精深及大国工匠精神			
学习方法	看：观看微课视频及相关课程资料； 思：认真思考任务单中的问题； 记：将心中的疑惑记下来			
课堂学习形式预告	课堂外		课堂内	
	1. 自主学习，搜集资料； 2. 整理收获，提出问题		1. 展示交流； 2. 协作探究； 3. 启发式教学； 4. 项目驱动； 5. 实习作业法	
本次学习任务	任务1：熟练掌握古建筑墙体类型； 任务2：熟练掌握墀头与埠头的作用； 任务3：熟练掌握按照材料划分的墙体种类			
疑点与问题	本次学习中还有哪些不明白的地方？请依次记录下来。			

总结评价	1. 你认为自己预习得怎么样？
	   A. 马马虎虎，只是为了完成任务
	   B. 态度认真，收获很多
	 2. 你对自己本次自主学习的评价：☆☆☆☆☆			

资讯单

学习领域	墙体构造		
学习情境	学习墙体构造的相关知识点		
学习任务	古建筑墙体类型、墀头与墀头的作用、按照材料划分的墙体种类	学时	10分钟
资讯方式	1. 教师讲解； 2. 互联网查询； 3. 同学交流		
资讯问题	1. 古建筑的墙体类型有哪些？ 2. 什么是墀头？ 3. 墀头的作用是什么？ 4. 按照材料划分的墙体种类有哪些？ 5. 通过墙体构造的学习内容谈中国古建筑中暗藏的民族智慧		
资讯引导	1. 微课视频； 2. PPT； 3. 课程相关视频资料		

计划单

学习领域	墙体构造		
学习情境	学习墙体构造的相关知识点		
学习任务	古建筑墙体类型、墀头与墀头的作用、按照材料划分的墙体种类	学时	10分钟
典型工作过程描述	教学情境导入—古建筑墙体类型—墀头与墀头的作用—按照材料划分的墙体类型—民族文化精神解读		
计划制订的方式	同学间分组讨论		

序号	工作步骤	注意事项
1	查看资料文件	
2	查询资料	
3	设计课程内容思维导图	

计划评价	班级		第___组	组长签字	
	教师签字		日期		
	评语：				

项目3 墙体构造 027

决策单

学习领域	墙体构造		
学习情境	学习墙体构造的相关知识点		
学习任务	古建筑墙体类型、墀头与墀头的作用、按照材料划分的墙体种类	学时	10分钟
典型工作过程描述	教学情境导入—古建筑墙体类型—墀头与墀头的作用—按照材料划分的墙体类型—民族文化精神解读		

学生为主体的课堂决策					
序号	课程目标	课程内容	教学活动	教学资料	综合评价
1					
2					
3					
4					
5					
6					
7					
8					
9					
10					

	班级		第___组	组长签字	
	教师签字		日期		
决策评价	评语:				

实施单

学习领域	墙体构造		
学习情境	学习墙体构造的相关知识点		
学习任务	古建筑墙体类型、墀头与墀头的作用、按照材料划分的墙体种类	学时	30分钟
典型工作过程描述	教学情境导入—古建筑墙体类型—墀头与墀头的作用—按照材料划分的墙体类型—民族文化精神解读		

序号	实施步骤	注意事项
1	教学情境导入	播放先导片
2	古建筑墙体类型	教学重点与难点
3	墀头与墀头的作用	熟练掌握
4	按照材料划分的墙体类型	教学重点与难点
5	结合课程思政	引入中华民族传统文化精神

实施说明:
1. 播放先导片;
2. 借助微课视频与PPT深入了解;
3. 通过了解古建筑墙体构造引入中华民族传统文化精神

	班级		第___组	组长签字	
	教师签字		日期		
实施评价	评语:				

项目3 墙体构造

检查单

学习领域	墙体构造			
学习情境	学习墙体构造的相关知识点			
学习任务	古建筑墙体类型、墀头与墀头的作用、按照材料划分的墙体种类		学时	15分钟
典型工作过程描述	教学情境导入—古建筑墙体类型—墀头与墀头的作用—按照材料划分的墙体类型—民族文化精神解读			
序号	检查项目	检查标准	学生自查	教师检查
1	资讯环节	获取相关信息情况		
2	计划环节	设计课程思维导图		
3	实施环节	绘制课程思维导图		
4	检查环节	各个环节逐一检查		
检查评价	班级		第___组	组长签字
	教师签字		日期	
	评语：			

评价单

学习领域	墙体构造			
学习情境	学习墙体构造的相关知识点			
学习任务	古建筑墙体类型、墀头与墀头的作用、按照材料划分的墙体种类		学时	15分钟
典型工作过程描述	教学情境导入—古建筑墙体类型—墀头与墀头的作用—按照材料划分的墙体类型—民族文化精神解读			
评价项目	评价子项目	学生自评	组内评价	教师评价
资讯环节	1. 听取教师讲解； 2. 互联网查询情况； 3. 同学交流情况			
计划环节	1. 查询资料情况； 2. 设计思维导图			
实施环节	1. 学习态度； 2. 课程内容熟练程度； 3. 课堂表现； 4. 作业完成情况及质量			
最终结果	综合情况			
评价	班级		第___组	组长签字
	教师签字		日期	
	评语：			

3.2　理论指导

3.2.1　古建筑墙体类型

墙体构造

中国的万里长城是古建筑墙体的代表。它不是一道单纯孤立的城墙，而是以城墙为主体，同大量的城、障、亭、标相结合的防御体系。古人用砖砌、石砌、砖石混合砌的方法砌筑城墙，下面让我们一起探寻古人藏在墙壁里的智慧。

墙体属于古建筑的中分部分，按照墙体在建筑中的位置划分为山墙、墀（chí）头、檐墙、槛墙、扇面墙（后金墙）、隔断墙等类型。

1. 山墙

山墙一般称为外横墙。沿建筑物短轴方向布置的墙即横墙，建筑物两端的横向外墙一般称为山墙。古建筑一般都有山墙，它的作用主要是将建筑与邻居的住宅隔开和防火（图3-1）。

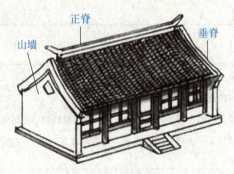

图3-1　古建筑山墙

山墙有3种形制，一种是人字形，这种造型简洁实用，修造成本也不高，民间多采用这种山墙（图3-2）；一种是锅耳形，其线条优美，变化大，实际上它是仿照古代的官帽形状修建的，取意前程远大，一般百姓的住宅中也常运用（图3-3）；另一种是波浪形，其造型起伏有致，讲究对称，起伏多为三级，实际是锅耳形山墙的变形，更像古代的官帽，百姓住宅中基本不用（图3-4）。

图3-2　人字形山墙

图3-3　锅耳形山墙

图3-4　波浪形山墙

2. 墀头

墀头由于位置特殊，远远看去像房屋昂扬的颈部，于是屋主都会用尽心思装饰。从每一处雕琢的痕迹中都可以看出它的精妙所在（图3-5、图3-6）。

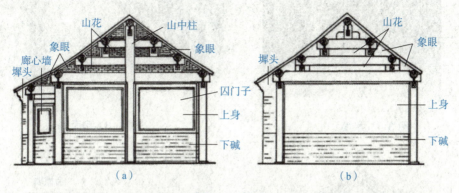

图3-5 墀头
（a）有山中柱；（b）无山中柱

图3-6 墀头装饰

3. 檐墙

处于檐柱和檐柱之间的墙叫作檐墙，即建筑物外部的纵墙，习惯上也称为外纵墙。其中，在建筑前檐的叫作"前檐墙"，在建筑后檐的叫作"后檐墙"。

4. 槛墙

槛墙是建筑前檐或后檐木装修风槛下的墙体（图3-7）。

5. 扇面墙

扇面墙又称为金内扇面墙，主要指前、后檐方向上金柱之间的墙体（图3-8）。

位于建筑物最外侧的称为檐柱，位于中轴线位置上的叫作山柱，位于檐柱与山柱之间的称为金柱（图3-9）。在这里，只要不是位于建筑物中轴线位置和最外侧位置的柱子都称为金柱，金柱之间的墙体就称为扇面墙。

6. 隔断墙

隔断墙又称为架山或夹山，是砌于前、后檐柱之间与山墙平行的内墙（图3-10）。

7. 廊墙

廊墙也叫作"廊心墙"，就是建筑的山墙里侧檐柱和金柱之间的墙（图3-11）。廊墙之上的中心位置还可以做装饰，形式也极为多样，有素心做法，还有绘画，甚至雕刻；

内容题材可以是几何纹、万字纹等吉祥纹样,也可以是花卉、鸟兽等(图 3-12)。

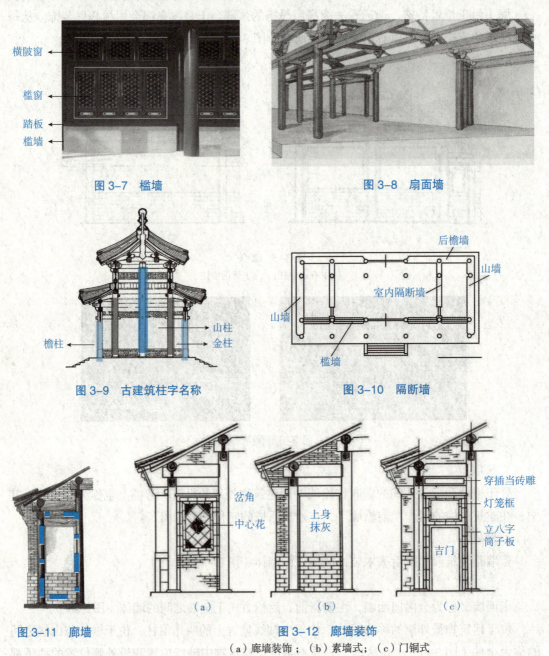

图 3-7 槛墙

图 3-8 扇面墙

图 3-9 古建筑柱子名称

图 3-10 隔断墙

图 3-11 廊墙

图 3-12 廊墙装饰
(a)廊墙装饰;(b)素墙式;(c)门铜式

3.2.2 古建筑墙体形式

按照材料划分,古建筑墙体主要分为土墙、砖墙、石墙。

1. 土墙

(1)夯土墙。夯土墙是我国最古老的墙体形式之一。土墙的隔热、隔声性能好,又有一定的承载能力,并可就地取材,施工也很简单,但易受自然侵蚀,特别是水浸后墙体的强度大大降低，因此古代筑墙时很注意选址和排水（图3-13）。

（2）三合土墙。三合土墙是由三合土夯筑而成的墙。三合土由石灰、黏土（或碎砖、碎石）和细砂组成。三合土经分层夯实，具有一定的强度和耐水性，可以将这种混合物夯打成最坚实的土墙。最典型的使用三合土墙的建筑就是福建土楼（图3-14）。

图3-13　夯土墙

图3-14　福建土楼的三合土墙

2. 砖墙

砖墙，即用青砖砌筑的墙。青砖因为经过烧制，所以硬度大，以其砌筑的墙壁比较坚固结实，不易毁坏。

我国很早就有使用砖的记载，但砖最初多用于墓室及塔的建造，而较少用在普通居住建筑上。直到明代，真正的砖砌墙才比较普遍（图3-15）。

（1）清水墙。凡是墙体表面不加粉刷、不加贴面材料的砖墙，都叫作清水墙。清水墙墙面上的砖缝多用泥沙或灰浆勾填，以防止雨水和风的侵入。勾缝后的墙体外表看起来更为整洁，虽然不华丽，但是显得朴实、干净（图3-16）。

图3-15　砖墙

（2）空心墙。空心墙具有明显节约材料的特点，非常经济。它的稳固性并不因此变差，有时候这样的空心墙还可以作为荷载墙。空心墙还具有良好的隔声、隔热性能，既经济又实用（图3-17）。

图3-16　清水墙

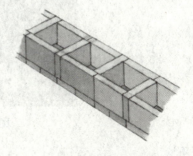

图3-17　空心墙

①漏砖墙。漏砖墙是花式砖墙的一种，即在墙洞处用砖砌成菱花或做出竹节等雕饰

（图3-18）。

图3-18　漏砖墙

②漏窗墙（图3-19）。漏窗墙是带有漏窗的墙体，常见于园林建筑中。这种墙体首先起隔断景区的作用，但园林的意境又要求它不能完全封闭，而是要隔而不断，因此这种带有漏窗的墙就被称为漏窗墙。

图3-19　漏窗墙

③砖花墙（图3-20）。砖花墙是指通过砖体凹凸不平的变化所形成的富有韵律美的表面图案的墙，其做法多样。

图3-20　砖花墙

3. 石墙

以石材砌筑的墙体称为石墙。石材是天然材料，有非常强的可塑性。其中最典型的石墙是虎皮石墙。

虎皮石墙是用在古建筑上的一种石墙，其用形状不规则的毛石砌筑，毛石之间用灰勾缝，灰缝与石块轮廓吻合（图3-21）。

从观感上，尤其远观，其色调、纹路的质感有些类似虎皮，因此工匠师傅们就约定俗成地把这种垒砌得像虎皮的石墙叫作虎皮石墙，并延续至今。虎皮石墙外观质朴，多用于寺庙、宅院等。

图3-21　虎皮石墙

3.2.3　古建筑墙体的作用

古建筑的墙体作用分为以下几点。

1. 承载作用

承载作用主要体现在砖木混合结构的房屋。由于后檐墙不设木柱屋架，所以荷载由前部木构架和后檐墙共同承担。

2. 围护与分隔室内空间作用

墙体通常将室内或分成左、中、右三部分，形成"一堂二内"的基本格局。

3. 防火作用

古建筑墙体材料多为砖、土坯、石等，其防火性能远高于木材。

4. 安全防御作用

古建筑墙体扩大建设可形成城墙、寨墙等，是防御性构筑物的集中体现，在古代战争中起着重要的防御作用。

5. 装饰作用

古建筑在墙体砌筑方面有严格的工艺要求，不同的墙面可以展现不同的艺术效果，如漏砖墙、漏窗墙及砖花墙。

3.3　中华民族传统文化精神解读

墙像一面镜子，不同时代的人通过它反观自己；墙像一位战士，巍然屹立，守护家园。一墙之隔，风景不同，墙内满庭景色，墙外百态众生。砖花墙砌筑形式的凹凸变化、韵律起伏形成了不同的图案变化效果，产生了不一样的感观体验。古人精湛的技艺与智慧从一砖一瓦间谱写出建造艺术的不朽传奇。

3.4　项目创新

通过墙体构造的学习内容写出对中国古建筑中所暗藏的民族智慧的感想，形式不限。

项目4 木构架构造

4.1 项目表单

学习任务单

学习领域	木构架构造				
学习情境	学习木构架构造的相关知识点				
学习任务	熟练掌握古建的抬梁式木构架、穿斗式木构架及井干式木构架构造相关知识点		学时	2学时（90分钟）	
教学目标	使学生充分了解古建筑的抬梁式木构架、穿斗式木构架及井干式木构架的结构设计				
技能目标	通过木构架构造进一步了解中国古建筑结构的精髓及中国优良的传统文化				
知识目标	掌握抬梁式木构架构造、穿斗式木构架构造、井干式木构架构造的相关知识点				
学习目标	1. 熟练掌握抬梁式木构架构造； 2. 熟练掌握穿斗式木构架构造； 3. 熟练掌握井干式木构架构造				
任务描述	熟练掌握木构架构造的相关知识点				
学时安排	情境导入	计划	决策	实施	检查	评价
时长	10分钟	10分钟	10分钟	30分钟	15分钟	15分钟
对学生的要求	1. 课前做好项目预习； 2. 做好课中总结与评价； 3. 通过古建筑墙体构造的学习内容谈建筑语言中的人文主义精神及中华民族传统文化				
参考资料	1. 微课视频； 2. PPT； 3. 课程相关视频资料				

项目 4　木构架构造　　037

项目前学习任务单

学习领域	木构架构造		用时	
达成目标	1. 观看微课视频，可以简单整理出课程内容思维导图； 2. 根据课程内容，感受中国古建筑文化的博大精深及大国工匠精神			
学习方法	看：观看微课视频及相关课程资料； 思：认真思考任务单中的问题； 记：将心中的疑惑记下来			
课堂学习形式预告	课堂外		课堂内	
	1. 自主学习，搜集资料； 2. 整理收获，提出问题		1. 展示交流； 2. 协作探究； 3. 启发式教学； 4. 项目驱动； 5. 实习作业法	
本次学习任务	任务1：熟练掌握抬梁式木构架构造； 任务2：熟练掌握穿斗式木构架构造； 任务3：熟练掌握井干式木构架构造			
疑点与问题	本次学习中还有哪些不明白的地方？请依次记录下来。			
总结评价	1. 你认为自己预习得怎么样？ 　A. 马马虎虎，只是为了完成任务 　B. 态度认真，收获很多 2. 你对自己本次自主学习的评价：☆☆☆☆☆			

资讯单

学习领域	木构架构造		
学习情境	学习木构架构造的相关知识点		
学习任务	熟练掌握古建的抬梁式木构架、穿斗式木构架及井干式木构架构造相关知识点	学时	10分钟
资讯方式	1. 教师讲解； 2. 互联网查询； 3. 同学交流		
资讯问题	1. 古建筑的木构架构造类型有哪些？ 2. 抬梁式木构架的含义及特点是什么？ 3. 穿斗式木构架的含义及特点是什么？ 4. 井干式木构架的含义及特点是什么？ 5. 通过古建筑木构架构造的学习内容谈建筑语言中的人文主义精神及中华民族传统文化		
资讯引导	1. 微课视频； 2. PPT； 3. 课程相关视频资料		

决策单

学习领域	木构架构造		
学习情境	学习木构架构造的相关知识点		
学习任务	熟练掌握古建的抬梁式木构架、穿斗式木构架及井干式木构架构造相关知识点	学时	10分钟
典型工作过程描述	教学情境导入—古建筑木构架类型—抬梁式木构架的含义及特点—穿斗式木构架的含义及特点—井干式木构架的含义及特点—民族文化精神解读		

学生为主体的课堂决策				
序号	课程目标	课程内容	教学活动	教学资料	综合评价
1					
2					
3					
4					
5					
6					
7					
8					
9					
10					

决策评价	班级		第___组	组长签字	
	教师签字		日期		
	评语：				

项目 4　木构架构造

实施单

学习领域	木构架构造		
学习情境	学习木构架构造的相关知识点		
学习任务	熟练掌握古建的抬梁式木构架、穿斗式木构架及井干式木构架构造相关知识点	学时	30 分钟
典型工作过程描述	教学情境导入—古建筑木构架类型—抬梁式木构架的含义及特点—穿斗式木构架的含义及特点—井干式木构架的含义及特点—民族文化精神解读		

序号	实施步骤	注意事项
1	教学情境导入	播放先导片
2	古建筑木构架类型	教学重点
3	抬梁式木构架的含义及特点	教学难点
4	穿斗式木构架的含义及特点	教学难点
5	井干式木构架的含义及特点	教学难点
6	结合课程思政	引入中华民族传统文化精神

实施说明：
1. 播放先导片；
2. 借助微课视频与 PPT 深入了解；
3. 通过了解古建筑木构架构造引入中华民族传统文化精神

	班级		第___组	组长签字	
	教师签字		日期		
实施评价	评语：				

检查单

学习领域	木构架构造		
学习情境	学习木构架构造的相关知识点		
学习任务	熟练掌握古建的抬梁式木构架、穿斗式木构架及井干式木构架构造相关知识点	学时	15分钟
典型工作过程描述	教学情境导入—古建筑木构架类型—抬梁式木构架的含义及特点—穿斗式木构架的含义及特点—井干式木构架的含义及特点—民族文化精神解读		

序号	检查项目	检查标准	学生自查	教师检查
1	资讯环节	获取相关信息情况		
2	计划环节	设计课程思维导图		
3	实施环节	绘制课程思维导图		
4	检查环节	各个环节逐一检查		

	班级		第___组	组长签字	
	教师签字		日期		
检查评价	评语：				

评价单

学习领域	木构架构造			
学习情境	学习木构架构造的相关知识点			
学习任务	熟练掌握古建的抬梁式木构架、穿斗式木构架及井干式木构架构造相关知识点		学时	15分钟
典型工作过程描述	教学情境导入—古建筑木构架类型—抬梁式木构架的含义及特点—穿斗式木构架的含义及特点—井干式木构架的含义及特点—民族文化精神解读			
评价项目	评价子项目	学生自评	组内评价	教师评价
资讯环节	1. 听取教师讲解； 2. 互联网查询情况； 3. 同学交流情况			
计划环节	1. 查询资料情况； 2. 设计思维导图			
实施环节	1. 学习态度； 2. 课程内容熟练程度； 3. 课堂表现； 4. 作业完成情况及质量			
最终结果	综合情况			
评价	班级		第___组	组长签字	
	教师签字		日期		
	评语：				

4.2 理论指导

4.2.1 抬梁式木构架

木构架构造

传统木构架结构主要分为 3 种形制：抬梁式木构架、穿斗式木构架及井干式木构架。抬梁式木构架多用于比较重要的木构建筑，穿斗式木构架一般用于民间建筑较多，井干式木构架多用于一些特殊的地区，如这个地区木材很多且没有其他材料。

抬梁式木构架是中国古建筑木构架的主要形式。这种木构架的特点是沿房屋进深方向架数层叠架的梁，梁逐层缩短，层间垫短柱或木块，最上层梁中间立小柱，形成三角形屋架。因为它的梁是层层叠加的，因此也被称作"叠梁式构架"（图 4-1、图 4-2）。

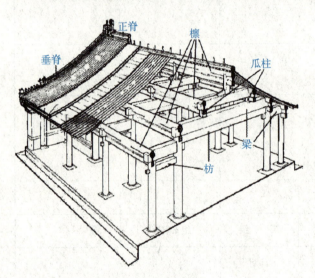

图 4-1 抬梁式木构架形制

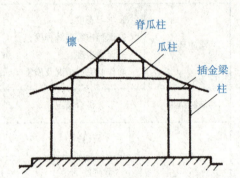

图 4-2 抬梁式木构架示意

抬梁式木构架的特点如下。

抬梁式木构架结构复杂，要求加工细致，但结实牢固，经久耐用，且内部有较大的使用空间，还能产生宏伟的气势，又可做出美观的造型（图 4-3）。

4.2.2 穿斗式木构架

穿斗式木构架是一种简洁、轻盈的木构架。穿斗式木构架同样是沿着房屋进深方向立柱，但是柱的排列较密，柱直接承受檩的质量，不用架空的抬梁，而是以数层"穿斗"贯穿各柱，组成一组组的木构架。穿斗式木构架被中国南方地区的建筑普遍采用（图 4-4）。

图 4-3 抬梁式木构架结构

图 4-4 穿斗式木构架形制

4.2.3 井干式木构架

井干式结构是一种不用立柱和大梁的房屋结构。这种结构以圆木或矩形、六角形木料平行向上层层叠置，在转角处木料端部交叉咬合，形成房屋四壁，形如古代井上的木围栏，再在左、右两侧壁上立矮柱承脊檩构成房屋（图4-5、图4-6）。

图 4-5 井干式木构架形制

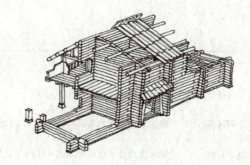

图 4-6 井干式木构架结构

4.3 中华民族传统文化精神解读

中国以木结构为主体的古建筑，在世界建筑之林中独树一帜，风格鲜明。它以木构架为骨，以砖石为体，结瓦为盖，油饰彩绘为衣，经历代能工巧匠精心设计，巧妙施工，潜心装饰，付诸心血和智慧建造而成，体现出东方古典建筑独有的艺术魅力及中国古建筑木结构的历史性、艺术性与科学性。

4.4 项目创新

结合前面所学内容将古建筑的结构从台基与地面、墙体构造到木构架构造绘制一张古建筑结构思维导图，形式不限。

项目 5　斗栱构造

5.1　项目表单

学习任务单

学习领域	斗栱构造					
学习情境	学习斗栱构造的相关知识点					
学习任务	熟练掌握古建的斗栱构造的含义与作用			学时	2学时（90分钟）	
教学目标	使学生充分了解古建筑的斗栱的含义与作用					
技能目标	通过木构架构造进一步了解中国古建筑结构的精髓及中国优良的传统文化					
知识目标	掌握古建筑斗栱构造的含义与作用					
学习目标	1. 熟练掌握斗栱的含义； 2. 熟练掌握斗栱的作用					
任务描述	熟练掌握木构架构造的相关知识点					
学时安排	情境导入	计划	决策	实施	检查	评价
时长	10分钟	10分钟	10分钟	30分钟	15分钟	15分钟
对学生的要求	1. 课前做好项目预习； 2. 做好课中总结与评价； 3. 通过古建筑斗栱构造的学习内容谈建筑语言中的人文主义精神及中华民族传统文化					
参考资料	1. 微课视频； 2. PPT； 3. 课程相关视频资料					

项目前学习任务单

学习领域	斗栱构造		用时	
达成目标	1. 观看微课视频，可以简单整理出课程内容思维导图； 2. 根据课程内容，感受中国古建筑文化的博大精深及大国工匠精神			
学习方法	看：观看微课视频及相关课程资料； 思：认真思考任务单中的问题； 记：将心中的疑惑记下来			
课堂学习形式预告	课堂外		课堂内	
	1. 自主学习，搜集资料； 2. 整理收获，提出问题		1. 展示交流； 2. 协作探究； 3. 启发式教学； 4. 项目驱动； 5. 实习作业法	
本次学习任务	任务1：熟练掌握斗栱的含义； 任务2：熟练掌握斗栱的作用			
疑点与问题	本次学习中还有哪些不明白的地方？请依次记录下来。 _____ _____ _____ _____			

总结评价	1. 你认为自己预习得怎么样？
	  A. 马马虎虎，只是为了完成任务
	  B. 态度认真，收获很多
	2. 你对自己本次自主学习的评价：☆☆☆☆☆			

资讯单

学习领域	斗栱构造		
学习情境	学习斗栱构造的相关知识点		
学习任务	熟练掌握古建筑的斗栱构造的含义与作用	学时	10分钟
资讯方式	1. 教师讲解； 2. 互联网查询； 3. 同学交流		
资讯问题	1. 古建筑的斗栱是什么？ 2. 古建筑的斗栱的作用是什么？ 3. 通过古建筑斗栱构造的学习内容谈建筑语言中的人文主义精神及中华民族传统文化		
资讯引导	1. 微课视频； 2. PPT； 3. 课程相关视频资料		

计划单

学习领域	斗栱构造		
学习情境	学习斗栱构造的相关知识点		
学习任务	熟练掌握古建筑的斗栱构造的含义与作用	学时	10分钟
典型工作过程描述	教学情境导入—古建筑斗栱的含义—古建筑斗栱的作用—民族文化精神解读		
计划制订的方式	同学间分组讨论		

序号	工作步骤	注意事项
1	查看资料文件	
2	查询资料	
3	设计课程内容思维导图	

计划评价	班级		第___组	组长签字	
	教师签字		日期		
	评语：				

决策单

学习领域	斗栱构造		
学习情境	学习斗栱构造的相关知识点		
学习任务	熟练掌握古建筑的斗栱构造的含义与作用	学时	10分钟
典型工作过程描述	教学情境导入—古建筑斗栱的含义—古建筑斗栱的作用—民族文化精神解读		

学生为主体的课堂决策

序号	课程目标	课程内容	教学活动	教学资料	综合评价
1					
2					
3					
4					
5					
6					
7					
8					
9					
10					

决策评价	班级		第___组	组长签字	
	教师签字		日期		
	评语：				

项目 5 斗栱构造　047

实施单

学习领域	斗栱构造		
学习情境	学习斗栱构造的相关知识点		
学习任务	熟练掌握古建的斗栱构造的含义与作用	学时	30 分钟
典型工作过程描述	教学情境导入—古建筑斗栱的含义—古建筑斗栱的作用—民族文化精神解读		
序号	实施步骤	注意事项	
1	教学情境导入	播放先导片	
2	斗栱的含义	熟练掌握	
3	结合课程思政	引入中华民族传统文化精神	

实施说明：
1. 播放先导片；
2. 借助微课视频与 PPT 深入了解；
3. 通过了解古建筑斗栱构造引入中华民族传统文化精神

实施评价	班级		第___组	组长签字	
	教师签字		日期		
	评语：				

检查单

学习领域	斗栱构造		
学习情境	学习斗栱构造的相关知识点		
学习任务	熟练掌握古建筑的斗栱构造的含义与作用	学时	15 分钟
典型工作过程描述	教学情境导入—古建筑斗栱的含义—古建筑斗栱的作用—民族文化精神解读		
序号	检查项目	检查标准	学生自查	教师检查
1	资讯环节	获取相关信息情况		
2	计划环节	设计课程思维导图		
3	实施环节	绘制课程思维导图		
4	检查环节	各个环节逐一检查		

检查评价	班级		第___组	组长签字	
	教师签字		日期		
	评语：				

评价单

学习领域	斗栱构造		
学习情境	学习斗栱构造的相关知识点		
学习任务	熟练掌握古建的斗栱构造的含义与作用	学时	15分钟
典型工作过程描述	教学情境导入—古建筑斗栱的含义—古建筑斗栱的作用—民族文化精神解读		

评价项目	评价子项目	学生自评	组内评价	教师评价
资讯环节	1. 听取教师讲解； 2. 互联网查询情况； 3. 同学交流情况			
计划环节	1. 查询资料情况； 2. 设计思维导图			
实施环节	1. 学习态度； 2. 课程内容熟练程度； 3. 课堂表现； 4. 作业完成情况及质量			
最终结果	综合情况			

评价	班级		第___组	组长签字	
	教师签字		日期		
	评语：				

5.2 理论指导

5.2.1 斗栱的含义

斗栱就是用来支撑房屋出檐的木构件（图5-1）。

图5-1 斗栱形制

斗栱构造

5.2.2 斗栱的作用

1. 增大距离作用

斗栱可把最外层的屋檐上的檩挑出一定距离，使建筑物出檐更加深远，造型更加优美、壮观。

2. 传递荷载作用

斗栱位于柱与梁之间，由屋面和上层构架传下来的荷载要通过斗栱传给柱子，再由柱传到基础，因此，斗栱起着承上启下，传递荷载的作用（图5-2）。

3. 装饰作用

斗栱构造精巧，造型美观，如盆景，似花篮，是很好的装饰性构件。后来的斗栱逐渐变为装饰，因为多采用砖墙，出檐变近许多，所以斗栱作用就弱化了。同时，斗栱也是区别建筑等级的标志。越高贵的建筑斗栱越复杂、繁复（图5-3）。

图5-2 斗栱的传递荷载作用

图5-3 斗栱的装饰作用

4. 抗震作用

榫卯结合是抗震的关键，斗栱把屋檐重量均匀地托住，起到了平衡稳定作用。中国古建筑"墙倒屋不塌"的关键就是木构件中的榫卯结构（图5-4）。

图 5-4　应县木塔

5.3　中华民族传统文化精神解读

在传统古建筑中，古代工匠们便用短木从柱子和梁上伸出，栱托着斗，斗托着栱，层层叠叠，组成中国古建筑中至美的元素。从皇家专用到走向民间，斗栱是一部关于中华建筑的史记。当你置身于斗栱之下时，总有到一种穿越古今的悸动，好像与千年前的古人同处于这片蓝天之下。

5.4　项目创新

结合前面课程所学内容，从"墙倒屋不塌"到古建筑的台基与地面、墙体构造、木构架构造到斗栱构造，谈一谈你对中国古建筑的印象与感想，可以用视频形式或文字形式表达。

项目 6　屋顶构造

6.1　项目表单

学习任务单

学习领域	屋顶构造		
学习情境	学习屋顶构造的相关知识点		
学习任务	熟练掌握古建筑的屋顶制式及屋脊名称的相关知识点	学时	2 学时（90 分钟）
教学目标	使学生充分了解古建筑的屋顶制式及屋脊名称		
技能目标	通过屋顶构造进一步了解中国古建筑结构的精髓及中国优良的传统文化		
知识目标	掌握古建筑的屋顶制式及屋脊名称的相关知识点		
学习目标	1. 熟练掌握古建筑的屋顶制式； 2. 熟练掌握古建筑屋脊名称		
任务描述	熟练掌握古建筑的屋顶制式及屋脊名称的相关知识点		

学时安排	情境导入	计划	决策	实施	检查	评价
时长	10 分钟	10 分钟	10 分钟	30 分钟	15 分钟	15 分钟

对学生的要求	1. 课前做好项目预习； 2. 做好课中总结与评价； 3. 通过古建筑屋顶构造的学习内容谈建筑语言中的人文主义精神及中华民族传统文化
参考资料	1. 微课视频； 2. PPT； 3. 课程相关视频资料

项目前学习任务单

学习领域	屋顶构造		用时	
达成目标	1. 观看微课视频，可以简单整理出课程内容思维导图； 2. 根据课程内容，感受中国古建筑文化的博大精深及大国工匠精神			
学习方法	看：观看微课视频及相关课程资料； 思：认真思考任务单中的问题； 记：将心中的疑惑记下来			
课堂学习形式预告	课堂外		课堂内	
	1. 自主学习，搜集资料； 2. 整理收获，提出问题		1. 展示交流； 2. 协作探究； 3. 启发式教学； 4. 项目驱动； 5. 实习作业法	
本次学习任务	任务1：熟练掌握古建筑的屋顶制式； 任务2：熟练掌握古建筑屋脊名称			
疑点与问题	本次学习中还有哪些不明白的地方？请依次记录下来。			
总结评价	1. 你认为自己预习得怎么样？ 　A. 马马虎虎，只是为了完成任务 　B. 态度认真，收获很多 2. 你对自己本次自主学习的评价：☆☆☆☆☆			

资讯单

学习领域	屋顶构造		
学习情境	学习屋顶构造的相关知识点		
学习任务	熟练掌握古建筑的屋顶制式及屋脊名称相关知识点	学时	10分钟
资讯方式	1. 教师讲解； 2. 互联网查询； 3. 同学交流		
资讯问题	1. 古建筑的屋顶构造形制有哪些？ 2. 古建筑屋顶各部分的名称是什么？ 3. 通过古建筑屋顶构造的学习内容谈建筑语言中的人文主义精神及中华民族传统文化		
资讯引导	1. 微课视频； 2. PPT； 3. 课程相关视频资料		

计划单

学习领域	屋顶构造		
学习情境	学习屋顶构造的相关知识点		
学习任务	熟练掌握古建筑屋顶制式及屋脊名称的相关知识点	学时	10 分钟
典型工作过程描述	教学情境导入—古建筑的屋顶制式—古建筑屋脊的名称—民族文化精神解读		
计划制订的方式	同学间分组讨论		

序号	工作步骤	注意事项
1	查看资料文件	
2	查询资料	
3	设计课程内容思维导图	

	班级		第___组	组长签字	
	教师签字		日期		
计划评价	评语：				

决策单

学习领域	屋顶构造		
学习情境	学习屋顶构造的相关知识点		
学习任务	熟练掌握古建筑的屋顶制式及屋脊名称的相关知识点	学时	10分钟
典型工作过程描述	教学情境导入—古建筑的屋顶制式—古建筑屋脊的名称—民族文化精神解读		

学生为主体的课堂决策

序号	课程目标	课程内容	教学活动	教学资料	综合评价
1					
2					
3					
4					
5					
6					
7					
8					
9					
10					

决策评价	班级		第___组	组长签字	
	教师签字		日期		
	评语：				

实施单

学习领域	屋顶构造		
学习情境	学习屋顶构造的相关知识点		
学习任务	熟练掌握古建筑的屋顶制式及屋脊名称的相关知识点	学时	30 分钟
典型工作过程描述	教学情境导入—古建筑的屋顶制式—古建筑屋脊的名称—民族文化精神解读		

序号	实施步骤	注意事项
1	教学情境导入	播放先导片
2	古建筑屋顶构造制式	教学重点与难点
3	古建筑屋脊的名称	教学重点与难点
4	结合课程思政	引入中华民族传统文化精神

实施说明：
1. 播放先导片；
2. 借助微课视频与 PPT 深入了解；
3. 通过了解古建筑屋顶构造引入中华民族传统文化精神

实施评价	班级		第___组	组长签字	
	教师签字		日期		
	评语：				

检查单

学习领域	屋顶构造		
学习情境	学习屋顶构造的相关知识点		
学习任务	熟练掌握古建筑的屋顶制式及屋脊名称的相关知识点	学时	15 分钟
典型工作过程描述	教学情境导入—古建筑的屋顶制式—古建筑屋脊的名称—民族文化精神解读		

序号	检查项目	检查标准	学生自查	教师检查
1	资讯环节	获取相关信息情况		
2	计划环节	设计课程思维导图		
3	实施环节	绘制课程思维导图		
4	检查环节	各个环节逐一检查		

检查评价	班级		第___组	组长签字	
	教师签字		日期		
	评语：				

评价单

学习领域	屋顶构造			
学习情境	学习屋顶构造的相关知识点			
学习任务	熟练掌握古建筑的屋顶制式及屋脊名称的相关知识点		学时	15 分钟
典型工作过程描述	教学情境导入—古建筑的屋顶制式—古建筑屋脊的名称—民族文化精神解读			
评价项目	评价子项目	学生自评	组内评价	教师评价
资讯环节	1. 听取教师讲解； 2. 互联网查询情况； 3. 同学交流情况			
计划环节	1. 查询资料情况； 2. 设计思维导图			
实施环节	1. 学习态度； 2. 课程内容熟练程度； 3. 课堂表现； 4. 作业完成情况及质量			
最终结果	综合情况			

评价	班级		第___组	组长签字	
	教师签字		日期		
	评语：				

6.2 理论指导

6.2.1 古建筑屋顶的形制

古建筑屋顶从形制上可分为硬山顶、悬山顶、庑殿顶及歇山顶。

1. 硬山顶

硬山顶有这样两个特点：屋顶为两面坡；靠山墙的屋檐不出挑（图6-1）。

2. 悬山顶

悬山顶也是两面坡，但是它与硬山顶最大的区别是山墙两边的屋檐就是挑出来的，挑出山墙以外的是悬山（图6-2）。

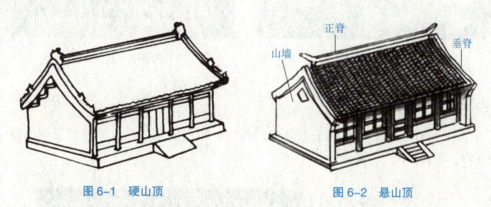

图6-1 硬山顶　　　　　　　　图6-2 悬山顶

3. 庑殿顶

庑殿顶最明显的标志就是有四面坡，并且是4种屋顶形制中规格最高的一种（图6-3）。

4. 歇山顶

歇山顶相当于在庑殿顶上切两刀，也就是在山墙上面的屋顶处切两刀，形成半坡面，同时还多出了4条戗脊（图6-4）。

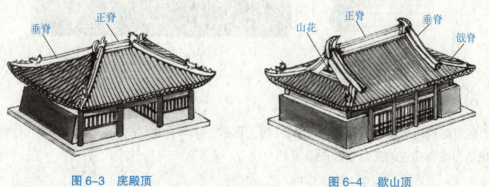

图6-3 庑殿顶　　　　　　　　图6-4 歇山顶

古建筑屋顶等级中，等级最高的是庑殿顶，其次是歇山顶，然后是悬山顶，最后是硬山顶。

6.2.2 重檐

有两层屋顶的形制在建筑语言中叫作重檐，重檐一般分为重檐庑殿顶与重檐歇山顶（图6-5、图6-6）。

图6-5 重檐庑殿顶

图6-6 重檐歇山顶

6.2.3 攒尖顶

屋顶上面有一个很可爱的类似塔尖的构件，叫作宝顶，有宝顶的屋顶叫作攒尖顶（图6-7、图6-8）。

图6-7 宝顶

图6-8 攒尖顶

6.2.4 屋顶上的屋脊名称

戗脊是与房屋的前檐呈45°的屋脊，最上面的一条叫作正脊；与正脊呈90°的叫作垂脊；山花是与垂脊相连的三角形（图6-9）。

图 6-9 古建筑屋脊名称

6.2.5 万字顶

万字顶的建筑是藏在圆明园之中的一座宫殿,寓意为万方安和(图 6-10、图 6-11)。

图 6-10 圆明园中的万字顶宫殿遗址

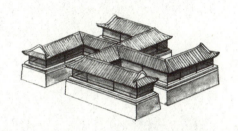

图 6-11 万字顶

6.3 中华民族传统文化精神解读

千百年来的中国屋顶到底隐藏了多少故事?它们在历史中静默,傲然于尘世,饱经风霜,受尽洗礼,却依然坚挺地给人们撑起一片深远的天空。站在屋顶的角度看中国,就是翻开一部石刻的史书,于千年风雨沧桑中,邂逅最明媚的风景。硬山顶、悬山顶、歇山顶及庑殿顶无一不在谱写中国古建筑的华美篇章。

6.4 项目创新

结合前面所学课程的内容,从古建筑的台基与地面、墙体构造、木构架构造、斗拱构造及今天所学习的屋顶构造,我们已经对整个古建筑房屋结构有了一定的认识。将这些古建筑的结构按照三段式整理出一张古建筑思维导图。

项目 7　木装修

7.1　项目表单

学习任务单

学习领域	木装修		
学习情境	学习木装修的相关知识点		
学习任务	熟练掌握古建筑内檐装修及外檐装修的相关知识点	学时	2学时（90分钟）
教学目标	使学生充分了解古建筑的内檐装修及外檐装修		
技能目标	通过木装修构造进一步了解中国古建筑结构的精髓及中国优良的传统文化		
知识目标	掌握古建筑内檐装修及外檐装修的相关知识点		
学习目标	1.熟练掌握古建筑内檐装修部分所包含的内容； 2.熟练掌握古建筑外檐装修部分所包含的内容； 3.花罩的分类		
任务描述	熟练掌握古建筑内檐装修及外檐装修的相关知识点		
学时安排	情境导入	计划	决策	实施	检查	评价
时长	10分钟	10分钟	10分钟	30分钟	15分钟	15分钟
对学生的要求	1.课前做好项目预习； 2.做好课中总结与评价； 3.通过古建筑木装修的学习内容谈建筑语言中的人文主义精神及中华民族传统文化		
参考资料	1.微课视频； 2.PPT； 3.课程相关视频资料		

项目前学习任务单

学习领域	木装修		用时	
达成目标	1. 观看微课视频，可以简单整理出课程内容思维导图； 2. 根据课程内容，感受中国古建筑文化的博大精深及大国工匠精神			
学习方法	看：观看微课视频及相关课程资料； 思：认真思考任务单中的问题； 记：将心中的疑惑记下来			
课堂学习形式预告	课堂外		课堂内	
	1. 自主学习，搜集资料； 2. 整理收获，提出问题		1. 展示交流； 2. 协作探究； 3. 启发式教学； 4. 项目驱动； 5. 实习作业法	
本次学习任务	任务 1：熟练掌握古建筑内檐装修部分所包含的内容； 任务 2：熟练掌握古建筑外檐装修部分所包含的内容； 任务 3：熟练掌握花罩的分类			
疑点与问题	本次学习中还有哪些不明白的地方？请依次记录下来。			
总结评价	1. 你认为自己预习得怎么样？ 　A. 马马虎虎，只是为了完成任务 　B. 态度认真，收获很多 2. 你对自己本次自主学习的评价：☆☆☆☆☆			

资讯单

学习领域	木装修		
学习情境	学习木装修的相关知识点		
学习任务	熟练掌握古建筑内檐装修及外檐装修的相关知识点	学时	10 分钟
资讯方式	1. 教师讲解； 2. 互联网查询； 3. 同学交流		
资讯问题	1. 古建筑内檐装修部分包含哪些内容？ 2. 古建筑外檐装修部分包含哪些内容？ 3. 花罩有哪些分类？ 4. 通过古建筑木装修的学习内容谈建筑语言中的人文主义精神及中华民族传统文化		
资讯引导	1. 微课视频； 2. PPT； 3. 课程相关视频资料		

计划单

学习领域	木装修		
学习情境	学习木装修的相关知识点		
学习任务	熟练掌握古建筑内檐装修及外檐装修的相关知识点	学时	10分钟
典型工作过程描述	教学情境导入—古建筑内檐装修—古建筑外檐装修—花罩的分类—民族文化精神解读		
计划制订的方式	同学间分组讨论		

序号	工作步骤	注意事项
1	查看资料文件	
2	查询资料	
3	设计课程内容思维导图	

计划评价	班级		第___组	组长签字	
	教师签字		日期		
	评语：				

决策单

学习领域	木装修		
学习情境	学习木装修的相关知识点		
学习任务	熟练掌握古建筑内檐装修及外檐装修的相关知识点	学时	10分钟
典型工作过程描述	教学情境导入—古建筑内檐装修—古建筑外檐装修—花罩的分类—民族文化精神解读		

学生为主体的课堂决策

序号	课程目标	课程内容	教学活动	教学资料	综合评价
1					
2					
3					
4					
5					
6					
7					
8					
9					
10					

决策评价	班级		第___组	组长签字	
	教师签字		日期		
	评语：				

实施单

学习领域	木装修		
学习情境	学习木装修的相关知识点		
学习任务	熟练掌握古建筑内檐装修及外檐装修的相关知识点	学时	30 分钟
典型工作过程描述	教学情境导入—古建筑内檐装修—古建筑外檐装修—花罩的分类—民族文化精神解读		
序号	实施步骤	注意事项	
1	教学情境导入	播放先导片	
2	古建筑内檐装修	教学重点与难点	
3	古建筑外檐装修	教学重点与难点	
4	结合课程思政	引入中华民族传统文化精神	

实施说明：
1. 播放先导片；
2. 借助微课视频与 PPT 进行深入了解；
3. 通过了解古建筑屋木装修引入中华民族传统文化精神

实施评价	班级		第___组	组长签字	
	教师签字		日期		
	评语：				

检查单

学习领域	木装修		
学习情境	学习木装修的相关知识点		
学习任务	熟练掌握古建筑内檐装修及外檐装修的相关知识点	学时	15 分钟
典型工作过程描述	教学情境导入—古建筑内檐装修—古建筑外檐装修—花罩的分类—民族文化精神解读		
序号	检查项目	检查标准	学生自查	教师检查
1	资讯环节	获取相关信息情况		
2	计划环节	设计课程思维导图		
3	实施环节	绘制课程思维导图		
4	检查环节	各个环节逐一检查		

检查评价	班级		第___组	组长签字	
	教师签字		日期		
	评语：				

评价单

学习领域	木装修			
学习情境	学习木装修的相关知识点			
学习任务	熟练掌握古建筑内檐装修及外檐装修的相关知识点	学时	15分钟	
典型工作过程描述	教学情境导入—古建筑内檐装修—古建筑外檐装修—花罩的分类—民族文化精神解读			
评价项目	评价子项目	学生自评	组内评价	教师评价
资讯环节	1. 听取教师讲解； 2. 互联网查询情况； 3. 同学交流情况			
计划环节	1. 查询资料情况； 2. 设计思维导图			
实施环节	1. 学习态度； 2. 课程内容熟练程度； 3. 课堂表现； 4. 作业完成情况及质量			
最终结果	综合情况			
评价	班级		第___组	组长签字
	教师签字		日期	
	评语：			

7.2 理论指导

7.2.1 古建筑外檐装修

木装修

在以木结构体系为主体的中国古建筑中,做柱、梁、枋、檩等主体木结构的工种,称为大木作。与此对应,做建筑门窗、顶棚、室内隔断等木结构部分的工种称为小木作。这种木工分工明确记载于宋代《营造法式》中,一直沿用至今。小木作到清代也称为木装修,并细分为外檐装修和内檐装修。外檐装修包括门、窗及外檐柱装饰;内檐装修包括花罩、藻井及木楼梯装饰。

1. 外檐装修——门

门是中国古建筑中最重要的一个组成部分,它不仅是人们出入的通道,而且是古建筑等级的象征,昭示着地位、财富、文化品格。

(1)实踏门。实踏门常用于宫殿、王府等较高等级的古建筑中,门上常装饰有金钉,为等级地位的象征(图7-1)。

(2)棋盘门。其门扇看起来好似棋盘,因此称为"棋盘门"(图7-2)。

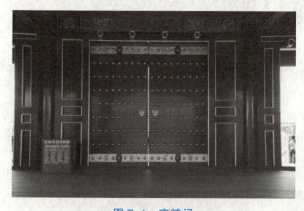

图7-1 实踏门

图7-2 棋盘门

(3)屋宇式大门。屋宇式大门是大门的主要形式,呈现为一座单独的房屋建筑形态,既是门又是屋。上自皇帝的宫室,下至普通百姓的住宅,屋宇式大门都有较为广泛的应用(图7-3)。

(4)门环。门环也被称作铺首,其作用是开关大门和叩门,为一种实用物件(图7-4)。

2. 外檐装修——窗

窗户犹如建筑的眼睛,通过它的"深眸"展示建筑的风韵,与建筑风格融为一体。同时,中国古建筑的窗户不仅实用还兼具美观效果,充满东方美学的色彩。

(1)直棂窗。直棂窗是棂条最为简单的一种窗户样式,棂条竖向排列,犹如栅栏(图7-5)。

（2）破子棂窗。破子棂窗的特点在于"破"字，它的窗棂是将方形断面的木料沿对角线斜破而成，即三角形断面的尖端朝外，平的一面朝内，以便于在窗内糊纸，用来遮挡风沙、抵御冷气等（图7-6）。

图7-3　屋宇式大门

图7-4　门环

图7-5　直棂窗

图7-6　破子棂窗

（3）槛窗。槛窗是一种形制较高级的窗户，是一种隔扇窗，窗扇上、下有转轴，可以向里和向外开关（图7-7）。

（4）支摘窗。支摘窗是一种可以支起、摘下的窗户，明清以来在普通住宅中常用。支摘窗一般分为上、下两段，上端可以推起支起，下端则可以摘下，这就是支摘窗名称的由来，同时也是它和槛窗最大的区别（图7-8）。

图7-7　槛窗

图7-8　支摘窗

（5）漏窗。漏窗是一类形式较为自由的窗户，但是它不能开启。漏窗有沟通内外景物的作用，通过漏窗可以看到另一边的景色，似通还隔，若隐若现。漏窗发展到后来，大多内置多姿多彩的图案，本身就是优美的景点（图7-9）。

（6）空窗。空窗与漏窗的区别：空窗只有窗洞而没有窗棂，在建筑中，如果空窗属于"虚"要素，那么漏窗则是"半实半虚"要素。空窗的设置可以使几个空间互相穿插渗透，将内外景致融为一体，又能增加景深，扩大空间。漏窗和空窗通常设置于园林建筑中，以获得深邃而优美的意境（图7-10）。

图7-9　漏窗　　　　　　　　　图7-10　空窗

（7）什锦窗。什锦窗的魅力不仅来自它的艺术造型，还来自窗套的色彩与装饰：内容丰富的砖雕题材，有自然花草类、吉祥图案类等；窗心部分花格排列规律有序的图案不仅给人以美的视觉感受，其蕴含的丰富寓意还给人以美的心理感受（图7-11）。

图7-11　什锦窗

3. 外檐柱装饰

外檐柱装饰的对象包括楣子、栏杆、雀替。

（1）楣子。楣子是安装于建筑檐柱间的兼有装饰和实用功能的装修。依位置不同，其分为吊（倒）挂楣子和坐凳楣子（图7-12、图7-13）。吊（倒）挂楣子安装于檐枋之下，有丰富和装点建筑立面的作用；坐凳楣子安装在檐下柱间，除有丰富立面的功能外，还可供人坐下休息。

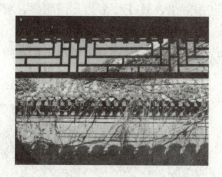 

图7-12　吊（倒）挂楣子　　　　图7-13　坐凳楣子

（2）栏杆。栏杆最初作为遮挡物，后来渐渐发展变化，式样丰富、雕刻精美，成了重要的装饰设施。在园林中，栏杆又起到隔景与连景的作用，功能似漏窗，而形象类花墙。栏杆中的柱子叫作望柱，两个柱子中间的部分叫作栏板（图7-14）。

图 7-14　栏杆中的望柱与栏板

（3）雀替。雀替原是放在柱子上端用来与柱子共同承受上部压力的物件，它的具体位置在梁与柱或枋与柱的交接处。雀替作为古建筑外檐装饰的构件体现着登峰造极的雕刻工艺，承载着丰富的文化内涵（图 7-15、图 7-16）。

图 7-15　雀替（一）

图 7-16　雀替（二）

7.2.2　古建筑内檐装修

1. 内檐装修——花罩

花罩在小木作中多用于室内装饰和隔断，分为很多种类，有飞罩、落地罩、栏杆罩、几腿罩。

飞罩使两柱门形成拱门状，但不落地。飞罩常用镂空的木格或雕花板做成，采用浮雕、透雕等手法表现出古拙、玲珑、清静、雅致的艺术效果，其花纹多为几何图案、缠交的动植物，或神话故事之类（图 7-17）。

落地罩是古建筑内檐装修木雕花罩的一种。凡从地上一直到梁（或枋）的花罩都可称为落地罩（图 7-18）。

图 7-17　飞罩

图 7-18　落地罩

栏杆罩就是中间带有栏杆的罩（图 7-19）。

几腿罩的两边有小的垂柱，而且柱头是不落地的，特别像桌、几的腿，因此叫作几腿罩（图 7-20）。

图 7-19　栏杆罩

图 7-20　几腿罩

2. 内檐装修——藻井

藻井是建筑物内部结构中最美的部分，可以说是抬头可见的惊艳。藻井是遮蔽建筑内顶部的构件，多饰以花纹、雕刻、彩画，故名藻井。藻井体现了巧夺天工的匠心、欲达天听的期盼，是中国人眼中最美的天宇（图 7-21、图 7-22）。

图 7-21　藻井（一）

图 7-22　藻井（二）

3. 内檐装修——木楼梯

古建筑中的木楼梯是作为楼层间垂直交通用的构件，主要用于楼层之间和高差较大时的交通联系（图 7-23）。

图 7-23 木楼梯

7.3 中华民族传统文化精神解读

从古建筑的木装修中,可以窥见东方韵律之美,每一处雕梁画栋、每一处幽幽棂窗都绽放出古人的匠心之美,巧夺天工的花罩及中国人眼中最美的天宇,让人心醉。琼楼玉宇是建筑音符的回荡,凝固着中华上下五千年的大国工匠精神。

7.4 项目创新

回顾本节课的内容,谈一谈在古建筑木装修中哪一部分给你留下了深刻的印象,将你的感想记录下来做成 PPT 进行汇报。

项目 8　彩画

8.1　项目表单

学习任务单

学习领域	彩画					
学习情境	学习彩画的相关知识点					
学习任务	熟练掌握古建筑彩画的等级、特点及作用的相关知识点			学时	2学时（90分钟）	
教学目标	使学生充分了解古建筑彩画的等级、特点及作用					
技能目标	通过古建筑彩画进一步了解中国古建筑结构的精髓及中国优良的传统文化					
知识目标	掌握古建筑彩画的等级、特点及作用的相关知识点					
学习目标	1.熟练掌握古建筑彩画的分类； 2.熟练掌握每种古建筑彩画的特点； 3.熟练掌握古建筑彩画的作用。					
任务描述	熟练掌握古建筑彩画的等级、特点及作用相关知识点					
学时安排	情境导入	计划	决策	实施	检查	评价
时长	10分钟	10分钟	10分钟	30分钟	15分钟	15分钟
对学生的要求	1.课前做好项目预习； 2.做好课中总结与评价； 3.通过古建筑彩画的学习内容谈建筑语言中的人文主义精神及中华民族传统文化					
参考资料	1.微课视频； 2.PPT； 3.课程相关视频资料					

项目前学习任务单

学习领域	彩画		用时	
达成目标	1. 观看微课视频，可以简单整理出课程内容思维导图； 2. 根据课程内容，感受中国古建筑文化的博大精深及大国工匠精神			
学习方法	看：观看微课视频及相关课程资料； 思：认真思考任务单中的问题； 记：将心中的疑惑记下来			
课堂学习形式预告	课堂外		课堂内	
	1. 自主学习，搜集资料； 2. 整理收获，提出问题		1. 展示交流； 2. 协作探究； 3. 启发式教学； 4. 项目驱动； 5. 实习作业法	
本次学习任务	任务 1：熟练掌握古建筑彩画的分类； 任务 2：熟练掌握每种古建筑彩画的特点； 任务 3：熟练掌握古建筑彩画的作用			
疑点与问题	本次学习中还有哪些不明白的地方？请依次记录下来。			
总结评价	1. 你认为自己预习得怎么样？ 　　A. 马马虎虎，只是为了完成任务 　　B. 态度认真，收获很多 2. 你对自己本次自主学习的评价：☆☆☆☆☆			

资讯单

学习领域	彩画		
学习情境	学习彩画的相关知识点		
学习任务	熟练掌握古建筑彩画的等级、特点及作用相关知识点	学时	10分钟
资讯方式	1. 教师讲解； 2. 互联网查询； 3. 同学交流		
资讯问题	1. 古建筑彩画分为哪几个等级？ 2. 古建筑彩画的特点有哪些？ 3. 古建筑彩画的作用是什么？ 4. 通过古建筑彩画的学习内容谈建筑语言中的人文主义精神及中华民族传统文化		
资讯引导	1. 微课视频； 2. PPT； 3. 课程相关视频资料		

计划单

学习领域	彩画		
学习情境	学习彩画的相关知识点		
学习任务	熟练掌握古建筑彩画的等级、特点及作用的相关知识点	学时	10分钟
典型工作过程描述	教学情境导入—古建筑彩画的等级—古建筑彩画的特点—古建筑彩画的作用—民族文化精神解读		
计划制订的方式	同学间分组讨论		
序号	工作步骤	注意事项	
1	查看资料文件		
2	查询资料		
3	设计课程内容思维导图		
计划评价	班级： 第___组 组长签字： 教师签字： 日期： 评语：		

决策单

学习领域	彩画		
学习情境	学习彩画的相关知识点		
学习任务	熟练掌握古建筑彩画的等级、特点及作用的相关知识点	学时	10分钟
典型工作过程描述	教学情境导入—古建筑彩画的等级—古建筑彩画的特点—古建筑彩画的作用—民族文化精神解读		

学生为主体的课堂决策

序号	课程目标	课程内容	教学活动	教学资料	综合评价
1					
2					
3					
4					
5					
6					
7					
8					
9					
10					

决策评价	班级： 第___组 组长签字： 教师签字： 日期： 评语：

实施单

学习领域	彩画		
学习情境	学习彩画的相关知识点		
学习任务	熟练掌握古建筑彩画的等级、特点及作用的相关知识点	学时	30分钟
典型工作过程描述	教学情境导入—古建筑彩画的等级—古建筑彩画的特点—古建筑彩画的作用—民族文化精神解读		

序号	实施步骤	注意事项
1	教学情境导入	播放先导片
2	古建筑彩画等级	教学重点与难点
3	古建筑彩画特点	教学重点与难点
4	古建筑彩画作用	教学重点与难点
5	结合课程思政	引入中华民族传统文化精神

实施说明:
1. 播放先导片;
2. 借助微课视频与PPT进行深入了解;
3. 通过了解古建筑屋木装修引入中华民族传统文化精神

实施评价	班级		第___组	组长签字	
	教师签字		日期		
	评语:				

检查单

学习领域	彩画		
学习情境	学习彩画的相关知识点		
学习任务	熟练掌握古建筑彩画的等级、特点及作用的相关知识点	学时	15分钟
典型工作过程描述	教学情境导入—古建筑彩画的等级—古建筑彩画的特点—古建筑彩画的作用—民族文化精神解读		

序号	检查项目	检查标准	学生自查	教师检查
1	资讯环节	获取相关信息情况		
2	计划环节	设计课程思维导图		
3	实施环节	绘制课程思维导图		
4	检查环节	各个环节逐一检查		

检查评价	班级		第___组	组长签字	
	教师签字		日期		
	评语:				

评价单

学习领域	彩画		
学习情境	学习彩画的相关知识点		
学习任务	熟练掌握古建筑彩画的等级、特点及作用的相关知识点	学时	15分钟
典型工作过程描述	教学情境导入—古建筑彩画的等级—古建筑彩画的特点—古建筑彩画的作用—民族文化精神解读		

评价项目	评价子项目	学生自评	组内评价	教师评价
资讯环节	1. 听取教师讲解； 2. 互联网查询情况； 3. 同学交流情况			
计划环节	1. 查询资料情况； 2. 设计思维导图			
实施环节	1. 学习态度； 2. 课程内容熟练程度； 3. 课堂表现； 4. 作业完成情况及质量			
最终结果	综合情况			

评价	班级		第___组	组长签字	
	教师签字		日期		
	评语：				

8.2 理论指导

彩画

8.2.1 彩画的等级

彩画在中国有悠久的历史，是古代传统建筑装饰中最突出的特点之一。它以独特的风格和独有的制作技术及其富丽堂皇的装饰艺术效果，给人留下了深刻印象，成为传统建筑艺术的精华而载入史册。

彩画在古建筑中也是分等级的，共分为3个等级，最高等级是和玺彩画，其次是旋子彩画，最后是苏式彩画。

8.2.2 彩画的特点

1. 和玺彩画

和玺彩画的主要特点是中间的画面由各种不同的龙或凤的图案组成，间补以花卉图案；画面两边用〈〉框住，并且沥粉贴金，金碧辉煌，十分壮丽。

沥粉贴金是和玺彩画最为突出的特点，和玺彩画中运用了大量的库金（库金就是我们平时说的金箔）。

和玺彩画主要用于紫禁城外朝的重要建筑及内廷中帝后居住等级较高的宫殿。太和殿、乾清宫、养心殿等宫殿多采用金龙和玺彩画（图8-1、图8-2）。

图8-1 和玺彩画（一）

图8-2 和玺彩画（二）

2. 旋子彩画

旋子彩画的等级次于和玺彩画。画面用简化形式的涡卷瓣旋花，有时也可画龙凤，两边用〈〉框起，可以贴金粉，也可以不贴金粉，一般用于次要宫殿或寺庙。旋子彩画多以青绿色为主（图8-3、图8-4）。

3. 苏式彩画

苏式彩画源于江南苏杭地区民间传统做法，俗称"苏州片"。其由图案和绘画两部分组成，主要用于古典园林建筑，给人以活泼、优雅、情趣与无限遐想。苏式彩画的题材丰富多样，包含花鸟、山水、人物、故事等（图8-5、图8-6）。

图8-3 旋子彩画（一）

图8-4 旋子彩画（二）

图8-5 苏式彩画（一）

图8-6 苏式彩画（二）

8.2.3 彩画的作用

彩画一方面有保护作用，另一方面起到装饰作用。人们原是为了使木结构防潮、防腐、防蛀，在其上刷颜料以保护木结构而形成彩画。宋代以后彩画已成为宫殿中不可缺少的装饰艺术，是我国古代建筑装饰中最突出的特点之一。

8.3　中华民族传统文化精神解读

彩画作为中国固有的建筑文化，在历史的洪流中熠熠生辉。它传达着中华民族悠久历史中的文化与精神，表达着不同的情感与寓意，体现了中华民族上下五千年文化的积淀。彩画随历史跌宕起伏，但贯彻始终，历久弥新。在漫长的历史轨迹中，彩画利用明丽的色彩与吉祥的纹路图式表现吉祥与教化世人的寓意，它作为一种独特的语言向我们描绘了先人的智慧、思想、审美，表达了先人对后人的期许与愿景，是我们了解、学习古文化的一种重要手段和载体。

8.4　项目创新

回顾本节课的内容，对于彩画的特点及彩画的修复过程谈一谈你的感想，并将你的感想记录下来做成PPT进行汇报。

项目9 宫廷建筑——紫禁城（故宫）

9.1 项目表单

学习任务单

学习领域	样式雷					
学习情境	学习故宫中"样式雷"的相关知识点					
学习任务	"样式雷"的形成、作用与分类			学时	2学时（90分钟）	
教学目标	使学生充分了解"样式雷"					
技能目标	通过"样式雷"进一步了解故宫中的建筑及中国传统文化中的大国工匠精神					
知识目标	掌握"样式雷"的概念、作用与分类					
学习目标	1. 了解"样式雷"的基本概念； 2. 了解"样式雷"的起源； 3. 熟练掌握"样式雷"的作用； 4. 熟练掌握"样式雷"的分类； 5. 了解"样式雷"的成就					
任务描述	熟练掌握"样式雷"的作用与分类					
学时安排	情境导入	计划	决策	实施	检查	评价
时长	10分钟	10分钟	10分钟	30分钟	15分钟	15分钟
对学生的要求	1. 课前做好项目预习； 2. 做好课中总结与评价； 3. 通过"样式雷"的学习内容谈大国工匠精神					
参考资料	1. 微课视频； 2. PPT； 3. 课程相关视频资料					

项目前学习任务单

学习领域	样式雷		用时	
达成目标	1. 观看微课视频，可以简单整理出课程内容思维导图； 2. 根据课程内容，感受中国文化的博大精深及大国工匠精神			
学习方法	看：观看微课视频及相关课程资料； 思：认真思考任务单中的问题； 记：将心中的疑惑记下来			
课堂学习形式预告	课堂外		课堂内	
	1. 自主学习，搜集资料； 2. 整理收获，提出问题		1. 展示交流； 2. 协作探究； 3. 启发式教学； 4. 项目驱动； 5. 实习作业法	
本次学习任务	任务1：了解"样式雷"的基本概念； 任务2：了解"样式雷"的起源； 任务3：熟练掌握"样式雷"的作用； 任务4：熟练掌握"样式雷"的分类； 任务5：了解"样式雷"的成就			
疑点与问题	本次学习中还有哪些不明白的地方？请依次记录下来。			
总结评价	1. 你认为自己预习得怎么样？ 　A. 马马虎虎，只是为了完成任务 　B. 态度认真，收获很多 2. 你对自己本次自主学习的评价：☆☆☆☆☆			

资讯单

学习领域	样式雷		
学习情境	学习故宫中"样式雷"的相关知识点		
学习任务	"样式雷"的形成、作用与分类	学时	10分钟
资讯方式	1. 教师讲解； 2. 互联网查询； 3. 同学交流		
资讯问题	1. 什么是"样式雷"？ 2. "样式雷"是如何起源的？ ★3. "样式雷"的作用是什么？ ★4. "样式雷"的分类包括哪些？ 5. "雷氏家族"的主要成就有哪些？ ★6. 通过"样式雷"的学习内容谈大国工匠精神		
资讯引导	1. 微课视频； 2. PPT； 3. 课程相关视频资料		

计划单

学习领域	样式雷		
学习情境	学习故宫中"样式雷"的相关知识点		
学习任务	"样式雷"的形成、作用与分类	学时	10分钟
典型工作过程描述	"样式雷"的起源—"样式雷"的作用—"样式雷"的分类		
计划制订的方式	同学间分组讨论		

序号	工作步骤	注意事项
1	查看资料文件	
2	查询资料	
3	设计课程内容思维导图	

计划评价	班级		第___组	组长签字	
	教师签字		日期		
	评语:				

决策单

学习领域	样式雷		
学习情境	学习故宫中"样式雷"的相关知识点		
学习任务	"样式雷"的形成、作用与分类	学时	10分钟
典型工作过程描述	"样式雷"的起源—"样式雷"的作用—"样式雷"的分类		

学生为主体的课堂决策

序号	课程目标	课程内容	教学活动	教学资料	综合评价
1					
2					
3					
4					
5					
6					
7					
8					
9					
10					

决策评价	班级		第___组	组长签字	
	教师签字		日期		
	评语:				

实施单

学习领域	样式雷		
学习情境	学习故宫中"样式雷"的相关知识点		
学习任务	"样式雷"的形成、作用与分类	学时	30 分钟
典型工作过程描述	教学情境导入—"样式雷"的起源—"样式雷"的作用—"样式雷"的分类		
序号	实施步骤	注意事项	
1	教学情境导入	播放先导片	
2	"样式雷"的起源	简单了解	
3	"样式雷"的作用	与成就一起理解	
4	"样式雷"的分类	教学重点与难点	
5	结合课程思政	引入"大国工匠精神"	

实施说明：
1. 播放先导片；
2. 借助微课视频与 PPT 深入了解"样式雷"的起源、作用与分类；
3. 引入大国工匠精神

实施评价	班级		第___组	组长签字	
	教师签字		日期		
	评语：				

检查单

学习领域	样式雷			
学习情境	学习故宫中"样式雷"的相关知识点			
学习任务	"样式雷"的形成、作用与分类		学时	15 分钟
典型工作过程描述	教学情境导入—"样式雷"的起源—"样式雷"的作用—"样式雷"的分类			
序号	检查项目	检查标准	学生自查	教师检查
1	资讯环节	获取相关信息情况		
2	计划环节	设计课程思维导图		
3	实施环节	绘制课程思维导图		
4	检查环节	各个环节逐一检查		

检查评价	班级		第___组	组长签字	
	教师签字		日期		
	评语：				

评价单

学习领域	样式雷		
学习情境	学习故宫中"样式雷"的相关知识点		
学习任务	"样式雷"的形成、作用与分类	学时	15分钟
典型工作过程描述	教学情境导入—"样式雷"的起源—"样式雷"的作用—"样式雷"的分类		

评价项目	评价子项目	学生自评	组内评价	教师评价
资讯环节	1. 听取教师讲解； 2. 互联网查询情况； 3. 同学交流情况			
计划环节	1. 查询资料情况； 2. 设计思维导图			
实施环节	1. 学习态度； 2. 课程内容熟练程度； 3. 课堂表现； 4. 作业完成情况及质量			
最终结果	综合情况			
评价	班级： 第___组 组长签字 教师签字 日期 评语：			

9.2 理论指导

9.2.1 "样式雷"的概念

"样式雷"是对清代 200 多年间主持皇家建筑设计的雷姓世家的誉称。

宫殿、皇陵、御苑等清代重要宫廷建筑和皇家工程，大多出自雷氏家族。"样式雷"可以说是赣派建筑中的瑰宝。在江右大地的山水毓秀文化浸润中，兴盛 200 余年的雷氏家族，其技艺为中国古建筑添上了浓墨重彩的一笔。

9.2.2 "样式雷"的起源

南方匠人雷发达来北京参加营造宫殿的工作，因为技术高超很快就被提升，担任设计工作（图 9-1）。从雷发达起直到清朝末年，主要的皇室建筑（如宫殿、皇陵、圆明园、颐和园等）都是由雷氏家族负责的，这个世袭的建筑师家族被称为"样式雷"。

200 年间，雷氏家族从雷发达到雷献彩共经历八代传承（图 9-2）。

图 9-1 雷发达

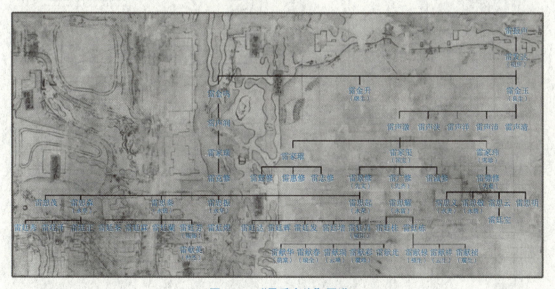

图 9-2 "雷氏家族"图谱

9.2.3 "样式雷"的分类

"样式雷"分为图档与烫样。

1. 图档

"样式雷"图档用水墨绘制，并在建筑图与地盘图边注明建筑尺寸及其他施工说明。图档按设计程序分为勘测选址图、建筑设计草图、进呈图样、施工进程图；按投影方式分为平面图、剖面图、立面图、大样图、透视图（图9-3～图9-5）。

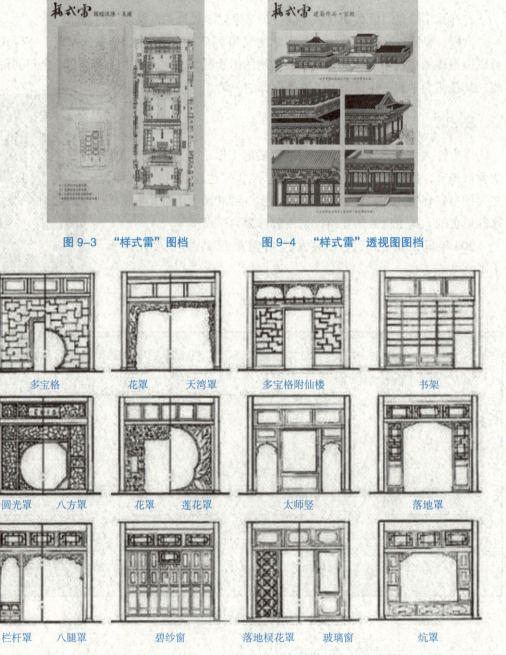

图9-3　"样式雷"图档　　　图9-4　"样式雷"透视图图档

图9-5　"样式雷"木装修图档

2. 烫样

烫样是用纸张、秫秸和木头加工制作成的模型图，因为最后用特制的小型烙铁将模型熨烫而成，所以被称为烫样。

烫样分为3个类别：第一类是全分样，即指组群建烫样，形成一个组群的模型（图9-6、图9-7）；第二类是分样，即单座建筑烫样、单体模型（图9-8、图9-9）；第三类是详样，指落地罩、屏风等（图9-10）。

图9-6 "样式雷"全分样（一）

长春宫烫样
《紫禁城》2009.8

图9-7 "样式雷"全分样（二）

图9-8 "样式雷"分样（一）

图9-9 "样式雷"分样（二）

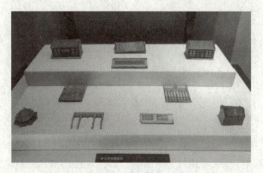

图9-10 "样式雷"详样

打开烫样的屋顶，可以看到建筑物内部的情况，如梁架结构、内檐彩画式样等。烫样上还贴有表示建筑各部尺寸的标签，如地安门烫样屋顶上的标签注明："地安门一座，面阔七间，宽十一丈四尺二寸，南北通进深三丈七尺六寸。明间面阔二丈二寸，次间面阔一丈七尺四寸，梢间面阔一丈五尺。檐柱高一丈八尺，径一尺八寸。中柱高二丈四尺二寸。

九檩歇山式屋顶，斗科单昂"。在烫样其他部位也有标签注明详细尺寸及构件名称。通过详细观察烫样，可以掌握地安门这座建筑从整体到细部的基本情况。

9.3　大国工匠精神解读

"建筑师"是今天的称谓，"工匠精神"也是今天的宣扬。尽管甚少有匠人能在青史上留名，但在人们的口耳相传中，他们演绎着鲁班那样神乎其神的传奇故事。

这些用纸张与木材营造的小模型，作为部分古建筑修缮或复建的重要依据，仍然在默默发挥着自己的作用。时至今日，"样式雷"家族及其作品依然在影响着我们、感动着我们。

9.4　项目创新

结合本节课的内容谈谈你对大国工匠精神的理解，把自己的想法与理解记录下来，形式不限。

项目 10　宫廷建筑——外朝三殿

10.1　项目表单

学习任务单

学习领域	外朝三殿					
学习情境	学习故宫中外朝三殿的相关知识点					
学习任务	外朝三殿的概况及功能			学时	2学时（90分钟）	
教学目标	使学生充分了解故宫的外朝三殿					
技能目标	通过外朝三殿进一步了解故宫中的建筑及中国传统文化					
知识目标	掌握外朝三殿的概况及功能					
学习目标	1. 熟练掌握外朝三殿的名称； 2. 熟练掌握太和殿内部装饰的意义及功能； 3. 熟练掌握中和殿内部装饰的意义及功能； 4. 熟练掌握保和殿内部装饰的意义及功能					
任务描述	熟练掌握外朝三殿的概况及功能					
学时安排	情境导入	计划	决策	实施	检查	评价
时长	10分钟	10分钟	10分钟	30分钟	15分钟	15分钟
对学生的要求	1. 课前做好项目预习； 2. 做好课中总结与评价； 3. 通过外朝三殿的学习内容进一步了解中华文化的博大精深					
参考资料	1. 微课视频； 2. PPT； 3. 课程相关视频资料					

项目前学习任务单

学习领域	外朝三殿		用时	
达成目标	1. 观看微课视频，可以简单整理出课程内容思维导图； 2. 根据课程内容，感受中国文化的博大精深			
学习方法	看：观看微课视频及相关课程资料； 思：认真思考任务单中的问题； 记：将心中的疑惑记下来			
课堂学习形式预告	课堂外		课堂内	
	1. 自主学习，搜集资料； 2. 整理收获，提出问题		1. 展示交流； 2. 协作探究； 3. 启发式教学； 4. 项目驱动； 5. 实习作业法	
本次学习任务	任务1：熟悉外朝三殿的名称； 任务2：了解太和殿内部装饰的意义及功能； 任务3：了解中和殿内部装饰的意义及功能； 任务4：了解保和殿内部装饰的意义及功能			
疑点与问题	本次学习中还有哪些不明白的地方？请依次记录下来。			
总结评价	1. 你认为自己预习得怎么样？ 　A. 马马虎虎，只是为了完成任务 　B. 态度认真，收获很多 2. 你对自己本次自主学习的评价：☆☆☆☆☆			

资讯单

学习领域	外朝三殿		
学习情境	学习故宫中外朝三殿的相关知识点		
学习任务	外朝三殿概况及功能	学时	10分钟
资讯方式	1. 教师讲解； 2. 互联网查询； 3. 同学交流		
资讯问题	★1. 故宫的外朝三殿都包括哪三大殿？ ★2. 太和殿的功能有哪些？ ★3. 太和殿的内部装饰包括哪些？分别有什么意义？ ★4. 中和殿的功能有哪些？ ★5. 保和殿的功能有哪些？		
资讯引导	1. 微课视频； 2. PPT； 3. 课程相关视频资料		

计划单

学习领域	外朝三殿		
学习情境	学习故宫中外朝三殿的相关知识点		
学习任务	外朝三殿的概况及功能	学时	10分钟
典型工作过程描述	外朝三殿的名称—太和殿内部装饰的意义及功能—中和殿内部装饰的意义及功能—保和殿内部装饰的意义及功能		
计划制订的方式	同学间分组讨论		

序号	工作步骤	注意事项
1	查看资料文件	
2	查询资料	
3	设计课程内容思维导图	

	班级		第___组	组长签字	
	教师签字		日期		
计划评价	评语：				

决策单

学习领域	外朝三殿		
学习情境	学习故宫中外朝三殿的相关知识点		
学习任务	外朝三殿的概况及功能	学时	10 分钟
典型工作过程描述	教学情境导入—外朝三殿的名称—太和殿内部装饰的意义及功能–中和殿内部装饰的意义及功能—保和殿内部装饰的意义及功能		

学生为主体的课堂决策

序号	课程目标	课程内容	教学活动	教学资料	综合评价
1					
2					
3					
4					
5					
6					
7					
8					
9					
10					

决策评价	班级		第___组	组长签字	
	教师签字		日期		
	评语：				

项目 10　宫廷建筑——外朝三殿　091

实施单

学习领域	外朝三殿			
学习情境	学习故宫中外朝三殿的相关知识点			
学习任务	外朝三殿的概况及功能		学时	30 分钟
典型工作过程描述	教学情境导入—外朝三殿的名称—太和殿内部装饰的意义及功能—中和殿内部装饰的意义及功能—保和殿内部装饰的意义及功能			
序号	实施步骤	注意事项		
1	教学情境导入	播放先导片		
2	外朝三殿包括哪些	熟练掌握		
3	太和殿的内部装饰意义及功能	明确功能，结合古建筑结构理解装饰意义，并熟练掌握		
4	中和殿功能	熟练掌握		
5	保和殿功能	熟练掌握		
6	结合课程思政	引入中华传统文化		

实施说明：
1. 播放先导片；
2. 借助微课视频与 PPT 深入了解"外朝三殿"的概况及功能；
3. 引入中华传统文化

实施评价	班级		第___组	组长签字	
	教师签字		日期		
	评语：				

检查单

学习领域	外朝三殿			
学习情境	学习故宫中外朝三殿的相关知识点			
学习任务	外朝三殿的概况及功能		学时	15 分钟
典型工作过程描述	教学情境导入—外朝三殿的名称—太和殿内部装饰的意义及功能—中和殿内部装饰的意义及功能—保和殿内部装饰的意义及功能			
序号	检查项目	检查标准	学生自查	教师检查
1	资讯环节	获取相关信息情况		
2	计划环节	设计课程思维导图		
3	实施环节	绘制课程思维导图		
4	检查环节	各个环节逐一检查		

检查评价	班级		第___组	组长签字	
	教师签字		日期		
	评语：				

评价单

学习领域	外朝三殿			
学习情境	学习故宫中外朝三殿的相关知识点			
学习任务	外朝三殿的概况及功能		学时	15分钟
典型工作过程描述	教学情境导入—外朝三殿的名称—太和殿内部装饰的意义及功能—中和殿内部装饰的意义及功能—保和殿内部装饰的意义及功能			
评价项目	评价子项目	学生自评	组内评价	教师评价
资讯环节	1. 听取教师讲解； 2. 互联网查询情况； 3. 同学交流情况			
计划环节	1. 查询资料情况； 2. 设计思维导图			
实施环节	1. 学习态度； 2. 课程内容熟练程度； 3. 课堂表现； 4. 作业完成情况及质量			
最终结果	综合情况			
评价	班级		第___组	组长签字
	教师签字		日期	
	评语：			

10.2 理论指导

10.2.1 故宫外朝三殿的名称

故宫外朝三殿

北京故宫内的建筑分为外朝和内廷两部分。外朝的中心为太和殿、中和殿、保和殿，统称三大殿，是国家举行盛大典礼的地方。内廷的中心是乾清宫、交泰殿及坤宁宫，统称后三宫，是皇帝和皇后居住的寝宫。

外朝是皇帝处理政事的地方，其中太和殿最为高大、辉煌，皇帝登基、皇帝大婚、册立皇后、命将出征等，都要在这里举行盛大典礼。

10.2.2 太和殿

1. 太和殿概况

太和殿是紫禁城内体量最大、等级最高的建筑物，是东方三大殿之一，中国现存最大的木结构大殿。它位于北京紫禁城南北主轴线的显要位置，面阔十一间，进深五间，长 64.24 m，宽 37 m，建筑面积为 2 377.00 m^2，高 26.92 m，连同台基通高 35.05 m，檐角安放 10 个走兽，前有宽阔的平台，称为丹陛，俗称月台，为紫禁城内规模最大的殿宇。

太和殿的一切构造都传达着一个信息——帝王是至高无上的，一切和谐都以他为中心（图 10-1）。

图 10-1 故宫太和殿

2. 太和殿的功能

有很多人认为，太和殿平时是用于皇帝上朝的，其实不是。太和殿其实是用来举行重大典礼的场所，实际使用次数很少。明清皇帝上朝的地方主要是太和门、乾清门（为御门听政）、乾清宫（有大事或重要的事情时皇帝召见大臣的地方），还有养心殿（清朝中后期皇帝居所），并不是平时所说的太和殿。明清两朝 24 个皇帝都在太和殿举行盛大典礼，如皇帝登基、皇帝大婚、册立皇后、命将出征，此外，每年万寿节、元旦、冬至三大节，皇帝在此接受文武官员的朝贺，并向王公大臣赐宴。

3. 太和殿内部装饰

太和殿内全部使用金砖铺地，但铺地的 4 718 块金砖不是用黄金制成的，而是用来自苏州的一种特制砖制成，因为敲击时能发出金属声，所以叫金砖。因此，太和殿又称为"金銮殿"（图10-2）。

太和殿内外装修极尽奢华，梁枋上绘有和玺彩画，门窗上部是菱花格纹，下部是浮雕云龙图案，接榫处安有镌刻龙纹的鎏金铜叶（图10-3）。

图10-2　太和殿内部金砖铺地

图10-3　太和殿内部装修奢华

10.2.3　中和殿

中和殿是皇帝去太和殿举行大典前稍事休息和演习礼仪的地方。皇帝在去太和殿之前先在此稍做停留，接受内阁大臣和礼部官员行礼，然后去太和殿举行仪式（图10-4）。

另外，皇帝祭祀天地和太庙之前，也要先在这里审阅写有祭文的"祝版"；在到中南海演讲前，也要在这里审视耕具。中和殿位于太和殿与保和殿之间，因此便有了中间意味。

"中和"二字意为凡事要做到不偏不倚，恰如其分才能使各方关系和顺，其意在于宣扬"中庸之道"。

10.2.4　保和殿

保和殿是每年除夕皇帝赐宴外藩王公的场所，明朝大典前皇帝常常在此更衣，册立皇后、皇太子时，皇帝在此殿受贺，保和殿也是举行科举考试殿试的地方（图10-5）。

图10-4　中和殿

图10-5　保和殿

10.2.5 三大殿的"千龙吐水"

三大殿中还隐藏着一个奇观,就是"千龙吐水"。"千龙"是指望柱下面伸出的千余个石雕龙头,每当雨天时雨水就从龙口中排出,使分流雨水的实用功能与建筑艺术的观赏功能有机地结合在一起(图10-6)。

"千龙吐水"的壮观景象出自太和、中和与保和三大殿矗立之上的三层台基。

图 10-6 "千龙吐水"奇观

10.3 中华民族传统文化精神解读

气势恢宏的三大殿可谓中国建筑历史上的艺术巅峰,每一处独具匠心的设计都融入了中华民族的工匠之魂、鲁班之魄。雕梁画栋间渗透着历史的印记,尽显民族恢宏的气势与魄力,吸天地之灵气,吮日月之精华,因此,弘扬民族精神是每一位中华儿女的职责所在。

10.4 项目创新

结合本节课的内容,写出自己的心得体会,并融入中华民族精神。

项目 11　宫廷建筑——屋脊十兽

11.1　项目表单

学习任务单

学习领域	屋脊十兽				
学习情境	学习故宫中屋脊十兽的相关知识点				
学习任务	屋脊十兽的名称及寓意		学时	2学时（90分钟）	
教学目标	使学生充分了解故宫中屋脊十兽的名称及寓意				
技能目标	通过屋脊十兽进一步了解故宫中的建筑及中国传统文化				
知识目标	掌握屋脊十兽的名称及寓意				
学习目标	1. 熟练掌握屋脊十兽的安放位置； 2. 熟练掌握屋脊十兽的名称； 3. 熟练掌握屋脊十兽的寓意； 4. 通过屋脊十兽进一步了解故宫中的建筑及中国传统文化				
任务描述	熟练掌握屋脊十兽的概况及功能				
学时安排	情境导入	计划	决策	实施	检查	评价
时长	10分钟	10分钟	10分钟	30分钟	15分钟	15分钟
对学生的要求	1. 课前做好项目预习； 2. 做好课中总结与评价； 3. 通过屋脊十兽的学习内容进一步了解中华文化的博大精深				
参考资料	1. 微课视频； 2. PPT； 3. 课程相关视频资料				

项目 11　宫廷建筑——屋脊十兽　097

项目前学习任务单

学习领域	屋脊十兽		用时	
达成目标	1. 观看微课视频，可以简单整理出课程内容思维导图（屋脊十兽的名称及寓意）； 2. 根据课程内容，感受中国文化的博大精深			
学习方法	看：观看微课视频及相关课程资料； 思：认真思考任务单中的问题； 记：将心中的疑惑记下来			
课堂学习形式预告	课堂外		课堂内	
	1. 自主学习，搜集资料； 2. 整理收获，提出问题		1. 展示交流； 2. 协作探究； 3. 启发式教学； 4. 项目驱动； 5. 实习作业法	
本次学习任务	任务 1：熟悉屋脊十兽的安放位置； 任务 2：熟练掌握屋脊十兽的名称； 任务 3：熟练掌握屋脊十兽的寓意； 任务 4：通过屋脊十兽的学习内容进一步了解中华文化的博大精深			
疑点与问题	本次学习中还有哪些不明白的地方？请依次记录下来。 _____ _____			
总结评价	1. 你认为自己预习得怎么样？ 　A. 马马虎虎，只是为了完成任务 　B. 态度认真，收获很多 2. 你对自己本次自主学习的评价：☆☆☆☆☆			

资讯单

学习领域	屋脊十兽		
学习情境	学习故宫中屋脊十兽的相关知识点		
学习任务	屋脊十兽的名称及寓意	学时	10 分钟
资讯方式	1. 教师讲解； 2. 互联网查询； 3. 同学交流		
资讯问题	★ 1. 屋脊十兽的安放位置在哪里？ ★ 2. 屋脊十兽为什么会有 11 个？ ★ 3. 屋脊十兽分别叫什么名字？ ★ 4. 屋脊十兽分别有什么寓意？		
资讯引导	1. 微课视频； 2. PPT； 3. 课程相关视频资料		

计划单

学习领域	屋脊十兽			
学习情境	学习故宫中屋脊十兽的相关知识点			
学习任务	屋脊十兽的名称及寓意		学时	10分钟
典型工作过程描述	屋脊十兽的名称—屋脊十兽的寓意			
计划制订的方式	同学间分组讨论			
序号	工作步骤		注意事项	
1	查看资料文件			
2	查询资料			
3	设计课程内容思维导图			
计划评价	班级		第___组	组长签字
	教师签字		日期	
	评语:			

决策单

学习领域	屋脊十兽			
学习情境	学习故宫中屋脊十兽的相关知识点			
学习任务	屋脊十兽的名称及寓意		学时	10分钟
典型工作过程描述	教学情境导入—屋脊十兽的名称—屋脊十兽的寓意			
学生为主体的课堂决策				
序号	课程目标	课程内容	教学活动	教学资料	综合评价
1					
2					
3					
4					
5					
6					
7					
8					
9					
10					
决策评价	班级		第___组	组长签字	
	教师签字		日期		
	评语:				

实施单

学习领域	屋脊十兽		
学习情境	学习故宫中屋脊十兽的相关知识点		
学习任务	屋脊十兽的名称及寓意	学时	30分钟
典型工作过程描述	教学情境导入—屋脊十兽的名称—屋脊十兽的寓意		
序号	实施步骤	注意事项	
1	教学情境导入	播放先导片	
2	屋脊十兽的名称	熟练掌握（思维导图）	
3	屋脊十兽的寓意	在熟练掌握名称的基础上，结合寓意进行理解（思维导图）	
4	通过屋脊十兽的学习内容进一步了解中华文化的博大精深	拓展内容	

实施说明：
1. 播放先导片；
2. 借助微课视频与PPT深入了解屋脊十兽的名称及寓意；
3. 引入中国传统文化

实施评价	班级		第___组	组长签字	
	教师签字		日期		
	评语：				

检查单

学习领域	屋脊十兽		
学习情境	学习故宫中屋脊十兽的相关知识点		
学习任务	屋脊十兽的名称及寓意	学时	15分钟
典型工作过程描述	教学情境导入—屋脊十兽的名称—屋脊十兽的寓意		
序号	检查项目	检查标准	学生自查	教师检查
1	资讯环节	获取相关信息情况		
2	计划环节	设计课程思维导图		
3	实施环节	绘制课程思维导图		
4	检查环节	各个环节逐一检查		

检查评价	班级		第___组	组长签字	
	教师签字		日期		
	评语：				

评价单

学习领域	屋脊十兽		
学习情境	学习故宫中屋脊十兽的相关知识点		
学习任务	屋脊十兽的名称及寓意	学时	15分钟
典型工作过程描述	教学情境导入—屋脊十兽的名称—屋脊十兽的寓意		

评价项目	评价子项目	学生自评	组内评价	教师评价
资讯环节	1. 听取教师讲解； 2. 互联网查询情况； 3. 同学交流情况			
计划环节	1. 查询资料情况； 2. 设计思维导图			
实施环节	1. 学习态度； 2. 课程内容熟练程度； 3. 课堂表现； 4. 作业完成情况及质量			
最终结果	综合情况			

评价	班级		第___组	组长签字	
	教师签字		日期		
	评语：				

11.2 理论指导

11.2.1 屋脊十兽概况

屋脊十兽是中国古代传统建筑中放置在房屋、宫殿等房脊上的雕塑作品。中国古建筑上的跑兽最多有 10 个，分布在房屋两端的垂脊上。屋脊十兽安放在故宫等级最高的太和殿之上。

屋脊十兽

11.2.2 屋脊十兽的名称

屋脊十兽由上至下依次为仙人、龙、凤、狮子、海马、天马、狎鱼、狻猊、獬豸、斗牛、行什。前面提到了屋脊十兽，这里为什么有 11 个呢？因为这里的仙人是后面 10 个神兽的首领。在古代，会根据主人的地位与级别在屋脊安放神兽，因此，代表最高等级皇权的太和殿才会有这 10 个神兽（图 11-1）。

图 11-1 屋脊十兽的名称

11.2.3 屋脊十兽的寓意

1. 仙人

仙人又称仙人骑凤，它的作用是固定垂脊下方的瓦片，它也标志着中国建筑美学的发展。它的寓意是骑凤飞行，逢凶化吉。同时它也在警告世人，再往前爬一点儿就会摔得粉身碎骨（图 11-2）。

2. 龙（鸱吻）

龙也称鸱吻，标志着至高无上的皇权。在古代，龙是避火灾的吉祥物。时至今日仍可见到"龙文化"在新建仿古建筑中的展示（图 11-3）。

图 11-2 仙人　　　　　图 11-3 龙（鸱吻）

3. 凤

凤在传统文化中用来比喻有圣德之人。所谓圣德，指的是西汉董仲舒所提出的"五常"，即仁、义、礼、智、信。同时凤也象征着皇后尊贵的地位（图 11-4）。

凤是从远古飞来的东方神鸟，它象征着美丽、吉祥，是中华民族文化的标志和精神象征。凤和龙一样，也不是现实中存在的动物，而是由生活在现实中的多种动物融合起来的神物。中国文化中一直有龙凤呈祥的美好象征。

4. 狮子

这里的狮子形象是经过汉化后的狮子，它的寓意是勇猛与威严。狮子作吼，群兽慑伏，代表勇猛、威严。狮子的样貌凶狠嚣张，针对一些妖魔鬼怪是有震慑作用的。狮子可以保佑家宅的安宁。石狮子就好像是守护神一样，可以保护家宅不被邪气入侵（图 11-5）。

图 11-4 凤　　　　　图 11-5 狮子

5. 海马

海马也称为落龙子，象征忠勇吉祥，其智慧与威德通天入海，畅达四方。海马入海入渊，逢凶化吉，在我国古代神话中是忠勇之兽。海马喜欢穿行于波涛之中，它的身上有振波及山行岩石图案，这些图案寓意为海水江崖，海马身上的图案也是古代吉祥纹样的代表（图 11-6）。

6. 天马

我们所熟知的成语"天马行空"中的天马指的就是屋脊十兽中的天马。天马可以日行千里，独来独往，它傲视群雄，开拓疆土，寓意逢凶化吉（图11-7）。

图11-6　海马　　　　　图11-7　天马

7. 狎鱼

狎鱼在海中，是执掌水族鱼类的统领，属于海中异兽。传说狎鱼和狻猊都是兴云作雨，灭火防灾的神。狎鱼与下雨谐音，据说它能喷出水柱，可以防火（图11-8）。

8. 狻猊

狻猊善于兴风作雨，虽生性凶猛，却心地善良，代表吉祥好运。它形似狮子，平生喜静不喜动，好坐，又喜欢烟火，传说能食虎豹，有率领百兽、护佑平安的寓意（图11-9）。

图11-8　狎鱼　　　　　图11-9　狻猊

9. 獬豸

獬豸头生一角，力大无穷，同时善断邪正，因此被视为法律的象征，清朝监察御史的官服上就绣有獬豸的图案。《异物志》中说："东北荒中有兽，名獬豸。一角，性忠，见人斗则触不直者，闻人论则咋不正者。"它能辨曲直，又有"神羊"之称，它是勇猛、公正的象征（图11-10）。

10. 斗牛

斗牛与虬龙、无角、狎鱼的作用相同，它们同为镇水兽，常被放置在水患发生之地。因此，它有镇邪、护宅之意（图 11-11）。

斗牛，牛角龙身，龙爪四指，常出现在明朝赏赐的三品官服上（称为斗牛服）。其形态是小一个等级的囚牛，是斗宿和牛宿间的神兽。

11. 行什

行什在 10 个神兽中排行第十，故而得名。它手中持有金刚宝杵，与传说中雷公的形象很相似，因此常被放置于屋顶防雷（图 11-12）。

同时，它也有另外一个寓意，即帮助人们守住秘密，舒缓情绪。

图 11-10 獬豸

图 11-11 斗牛

图 11-12 行什

11.3 中华民族传统文化精神解读

每一尊神兽都是古人的智慧与创造力的结晶，我们不得不惊叹于这些登峰造极的艺术形态，它们是凝聚着悠久的历史文化积淀与民族精神智慧的艺术瑰宝。通过对屋脊十兽的了解，我们领略了中华民族文化的博大精深。我们要将民族悠久的历史、优良的文化传统和道德修养发扬光大。

11.4 项目创新

回顾本节课中屋脊十兽的知识点及相关的历史元素，如海马身上的吉祥纹样，课后完成 PPT 总结汇报作业。

项目 12　宫廷建筑——故宫的建筑色彩及象征

12.1　项目表单

学习任务单

学习领域	故宫的建筑色彩及象征					
学习情境	学习故宫的建筑色彩及象征的相关知识点					
学习任务	熟练掌握故宫的建筑色彩分布及象征意义			学时	2学时（90分钟）	
教学目标	使学生充分了解故宫的建筑色彩及象征意义					
技能目标	通过故宫的建筑色彩及象征意义进一步了解故宫中的建筑及中国传统文化					
知识目标	故宫的建筑色彩及象征意义					
学习目标	1.熟练掌握故宫中的主要色彩分布； 2.熟练掌握每种颜色的象征意义； 3.通过故宫中的建筑色彩及象征意义进一步了解故宫中的建筑以及中国传统文化精髓					
任务描述	熟练掌握故宫的建筑色彩及象征意义					
学时安排	情境导入	计划	决策	实施	检查	评价
时长	10分钟	10分钟	10分钟	30分钟	15分钟	15分钟
对学生的要求	1.课前做好项目预习； 2.做好课中总结与评价； 3.通过故宫的建筑色彩及象征意义的学习内容进一步了解中国古建筑历史及中国传统文化精髓					
参考资料	1.微课视频； 2.PPT； 3.课程相关视频资料					

项目前学习任务单

学习领域	故宫的建筑色彩及象征	用时	
达成目标	1. 观看微课视频，可以简单整理出课程内容思维导图（故宫的建筑色彩分布及象征意义）； 2. 根据课程内容，感受中国传统文化的魅力		
学习方法	看：观看微课视频及相关课程资料； 思：认真思考任务单中的问题； 记：将心中的疑惑记下来		
课堂学习形式预告	课堂外	课堂内	
	1. 自主学习，搜集资料； 2. 整理收获，提出问题	1. 展示交流； 2. 协作探究； 3. 启发式教学； 4. 项目驱动； 5. 实习作业法	
本次学习任务	任务1：了解故宫中建筑色彩的分布； 任务2：了解故宫中建筑色彩的象征意义； 任务3：通过故宫中的建筑色彩及象征意义进一步了解故宫中的建筑以及中国传统文化精髓		
疑点与问题	本次学习中还有哪些不明白的地方？请依次记录下来。		
总结评价	1. 你认为自己预习得怎么样？ 　A. 马马虎虎，只是为了完成任务 　B. 态度认真，收获很多 2. 你对自己本次自主学习的评价：☆☆☆☆☆		

项目 12 宫廷建筑——故宫的建筑色彩及象征

资讯单

学习领域	故宫的建筑色彩及象征		
学习情境	学习故宫的建筑色彩及象征的相关知识点		
学习任务	熟练掌握故宫的建筑色彩分布及象征意义	学时	10分钟
资讯方式	1. 教师讲解； 2. 互联网查询； 3. 同学交流		
资讯问题	★ 1. 故宫中的建筑色彩是怎样分布的？ ★ 2. 故宫中的建筑色彩主要有哪些？从五个方面进行解读。 ★ 3. 故宫中建筑色彩所蕴含的意义是什么？ ★ 4. 通过对建筑色彩的认知了解中国古建筑与中国传统文化		
资讯引导	1. 微课视频； 2. PPT； 3. 课程相关视频资料		

计划单

学习领域	故宫的建筑色彩及象征		
学习情境	学习故宫的建筑色彩及象征的相关知识点		
学习任务	熟练掌握故宫的建筑色彩分布及象征意义	学时	10分钟
典型工作过程描述	教学情境导入—故宫的建筑色彩分布—色彩的象征意义		
计划制订的方式	同学间分组讨论		

序号	工作步骤	注意事项
1	查看资料文件	
2	查询资料	
3	设计课程内容思维导图	

计划评价	班级		第___组	组长签字	
	教师签字		日期		
	评语：				

决策单

学习领域	故宫的建筑色彩及象征			
学习情境	学习故宫的建筑色彩及象征的相关知识点			
学习任务	熟练掌握故宫的建筑色彩分布及象征意义		学时	10分钟
典型工作过程描述	教学情境导入—故宫的建筑色彩分布（瓦片、屋檐、柱架与墙体、台基与栏板及地面）—色彩的象征意义			

学生为主体的课堂决策

序号	课程目标	课程内容	教学活动	教学资料	综合评价
1					
2					
3					
4					
5					
6					
7					
8					
9					
10					

决策评价	班级		第___组	组长签字	
	教师签字		日期		
	评语：				

项目 12　宫廷建筑——故宫的建筑色彩及象征

实施单

学习领域	故宫的建筑色彩及象征		
学习情境	学习故宫的建筑色彩及象征的相关知识点		
学习任务	熟练掌握故宫的建筑色彩分布及象征意义	学时	30 分钟
典型工作过程描述	教学情境导入—故宫的建筑色彩分布（瓦片、屋檐、柱架与墙体、台基与栏板及地面）—色彩的象征意义		

序号	实施步骤	注意事项
1	教学情境导入	播放先导片
2	故宫的建筑色彩分布	熟练掌握（思维导图）
3	故宫建筑色彩的象征意义	在熟练掌握色彩分布的基础上，结合象征意义进行理解（思维导图）
4	通过对建筑色彩的认知了解中国古建筑与中国传统文化	拓展内容

实施说明：
1. 播放先导片；
2. 借助微课视频与 PPT 深入了解故宫的建筑色彩分布及象征意义；
3. 引入中国传统文化

实施评价	班级		第___组	组长签字	
	教师签字		日期		
	评语：				

检查单

学习领域	故宫的建筑色彩及象征		
学习情境	学习故宫的建筑色彩及象征的相关知识点		
学习任务	熟练掌握故宫的建筑色彩分布及象征意义	学时	15 分钟
典型工作过程描述	教学情境导入—故宫的建筑色彩分布（瓦片、屋檐、柱架与墙体、台基与栏板及地面）—色彩的象征意义		

序号	检查项目	检查标准	学生自查	教师检查
1	资讯环节	获取相关信息情况		
2	计划环节	设计课程思维导图		
3	实施环节	绘制课程思维导图		
4	检查环节	各个环节逐一检查		

检查评价	班级		第___组	组长签字	
	教师签字		日期		
	评语：				

评价单

学习领域	故宫的建筑色彩及象征		
学习情境	学习故宫的建筑色彩及象征的相关知识点		
学习任务	熟练掌握故宫的建筑色彩分布及象征意义	学时	15分钟
典型工作过程描述	教学情境导入—故宫的建筑色彩分布（瓦片、屋檐、柱架与墙体、台基与栏板及地面）—色彩的象征意义		

评价项目	评价子项目	学生自评	组内评价	教师评价
资讯环节	1. 听取教师讲解； 2. 互联网查询情况； 3. 同学交流情况			
计划环节	1. 查询资料情况； 2. 设计思维导图			
实施环节	1. 学习态度； 2. 课程内容熟练程度； 3. 课堂表现； 4. 作业完成情况及质量			
最终结果	综合情况			

评价	班级		第___组	组长签字	
	教师签字		日期		
	评语：				

12.2 理论指导

故宫的建筑色彩及象征

12.2.1 故宫中的色彩初印象

很多人对故宫建筑色彩的印象是红墙、黄瓦、朱门金钉，屋顶下偶见青绿色调的彩画装饰（图12-1～图12-4）。

图12-1 故宫中的红墙

图12-2 故宫中的黄瓦

图12-3 故宫中的朱门金钉

图12-4 故宫中的彩画装饰

12.2.2 故宫中的色彩美学解析及象征意义

下面分别从5个层面来解读故宫的建筑色彩，即瓦片、屋檐、柱架和墙体、台基和栏板、地面。

1. 瓦片

处于故宫中轴线位置的主要宫殿覆以黄色的瓦片，其东面的南三所覆以绿色的瓦片，西面的东西十二宫覆以金黄色的瓦片，南面的午门承台覆以红色瓦片，北面的文渊阁覆以黑色的瓦片（图12-5）。

（1）黄色琉璃瓦片。《周易·坤》里有"天玄而地黄"之语，意思是宇宙是高深莫测的，并孕育

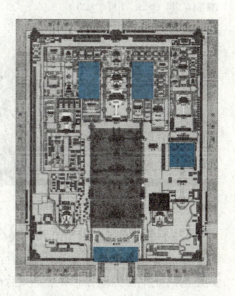

图12-5 故宫瓦片颜色分布示意

着大地。由于中华传统文化中大地是黄色的，所以用黄色来代表地。另外，土地是国家的象征，因此黄色也代表皇帝的权力。

《尚书》中有阴阳五行的说法，即金、木、水、火、土。其中黄色对应"土"，同时黄色也代表轻盈、活泼。

太和殿中的两层屋顶共有103片瓦片，每片瓦片都是工样瓦，质量为3.5～4 kg，这么重的屋顶正因为有了黄色的瓦片才显得不那么厚重，反而显得轻盈、活泼。象征皇权的黄色琉璃瓦也称为黄金外衣（图12-6、图12-7）。

图12-6 黄色琉璃瓦片

图12-7 覆盖黄色琉璃瓦片的建筑

平民百姓住宅的屋顶是不允许使用黄色瓦片的，一般使用黑色的瓦片，也称为布瓦，因此古时的百姓也称为布衣。

（2）绿色琉璃瓦片。绿色的瓦片位于东部，东部为太阳升起的地方。在阴阳五行中，东方属木，为青色，主生长。因此，皇子的居所在东方，覆以绿色琉璃瓦片寓意皇子可以茁壮成长。

故宫中的东部区域主要是皇子学习、生活的地方，这样"南三所"的建筑就显得与众不同（图12-8、图12-9）。

图12-8 绿色琉璃瓦片

图12-9 覆盖绿色琉璃瓦片的南三所

（3）金黄色琉璃瓦片。西部是太阳落山的地方，同时金黄色寓意金秋。金秋时节，万物都会有丰硕的成果。在阴阳五行中西方属金，因此西区在功能区域上的划分是供皇太后与后妃生活的地方，对于皇太后而言在这里可以安度晚年、一生圆满；对于妃子而言，寓意为希望她们可以多为皇帝生儿育女，以利于皇家子孙繁茂、多子多福（图12-10、图12-11）。

图12-10　覆盖以金黄色琉璃瓦片的后宫

图12-11　故宫中金黄色琉璃瓦片

（4）红色午门。南部主要指午门，午门的屋顶覆盖黄色琉璃瓦片，但它的承台是红色的。在阴阳五行中，南方属火，为红色。高高的承台表面饰以红色，既显得威严与庄重，又衬托了午门城楼的高大与雄伟（图12-12）。

（5）黑色琉璃瓦片。覆盖黑色琉璃瓦片的建筑叫作文渊阁，它的功能是藏书，也是皇帝的私人图书馆。

书籍最怕的就是火，俗话说水火不容，水就是用来克火的。

阴阳五行中黑色属水，因此文渊阁屋顶上覆盖黑色琉璃瓦片，寓意远离火灾（图12-13）。

图12-12　故宫午门

图12-13　覆盖黑色琉璃瓦片的文渊阁

2. 屋檐

屋檐主要是指额枋与斗栱，它们的颜色是青绿色。绿色使人想到大自然中的绿树和绿草，给人以春回大地、一片生机之感；蓝色使人联想到天空和海洋，也使人感到天地之广阔、大自然之宁静。

由于屋檐往外挑出，所以在梁枋下部及斗栱部位会出现阴影。冷色调在阴影中显得空气感强，轻盈而又遥远，使厚重的屋顶给人以轻松的感觉，而且增强了建筑的高度感和空间感，有扩大空间的感觉（图12-14）。

3. 柱架和墙体

故宫中柱架和墙体的颜色为红色，红色是紫禁城古建筑的主色之一。红色给人以希望和满足，自古以来，我国民间将红色当作喜庆的颜色。明清朝廷规定，凡专送皇帝的奏章必须为红色，说明帝王对红色的喜爱。

从功能上说，墙体对建筑起到围护作用，柱子是支持建筑屋顶的重要构件。可以看出，两种构件均能对建筑起到防御、保护的作用。其颜色为红色，有利于体现阳刚之气，有护卫皇家建筑之意（图12-15）。

图12-14　故宫中额枋与斗栱的色彩

图12-15　故宫中的红色柱架和墙体

4. 台基和栏板

台基和栏板为白色，白色是高雅、纯洁与尊贵的象征。由于紫禁城古建筑的台基、栏板和望柱上有精美的龙凤纹雕刻，所以采用洁白的汉白玉材料，这有利于突出建筑本身的高贵。

同时，白色的台基与黄色的屋顶、红色的柱子形成鲜明的对比，可显示出建筑的壮丽与高雅（图12-16）。

图12-16　故宫的台基和栏板

5. 地面

紫禁城古建筑中，室外地面为灰色，室内地面多为灰黑色（灰色是砖的本色，黑色为地面铺墁快完成时在表面泼洒的黑矾水的颜色）（图12-17、图12-18）。

图12-17　故宫中的室外地面

图12-18　故宫中的室内地面

从位置及功能的角度说，宫殿建筑地面不宜采用亮丽的色彩，因此采用低调的灰黑色，这种颜色位于各种色调中间，它无声地与各种色调结合，并融合于各种色调，形成了很好的补色效果。同时，与台基和栏板的白色相比，灰黑色与白色形成鲜明的色差，使同样为中间色调的白色获得了生命。

松弛的黑色和周密的白色使紫禁城的色彩趋于完美。

12.3　中华民族传统文化精神解读

紫禁城古建筑的色彩不仅体现了紫禁城古建筑本身的文化和艺术水平，更能反映出我国古代工匠的思维智慧。

可以说，建筑这一符号在一定程度上承载着一个国家或民族的文化，在中华上下五千年的历史文明长河中，不乏建筑这一重要角色的身影，而色彩作为建筑的一个重要组成元素，既能引起视觉上的种种反馈，又具有震撼的渲染力。

北京故宫的黄色琉璃瓦顶金碧辉煌，朱红色的柱子与门窗、檐下处于阴影部位的青绿色略点金的建筑彩画，在白色台基的衬托下，使建筑物各部分的轮廓更加鲜明，色彩别具一格，从而使建筑物更加富丽堂皇。各种不同的颜色是封建社会等级制度的体现，也代表了封建礼教在建筑上的内化。

12.4　项目创新

回顾本节课的知识点，根据所学内容画出思维导图。

项目 13 小宅门——四合院

13.1 项目表单

学习任务单

学习领域	四合院					
学习情境	学习四合院的相关知识点					
学习任务	熟练掌握四合院的分类、宅门及影壁制式、四合院各部分名称			学时	2学时（90分钟）	
教学目标	使学生充分了解四合院的分类、宅门及影壁制式、四合院各组成部分名称					
技能目标	通过四合院进一步了解中国古建筑结构的精髓及中国优良的传统文化					
知识目标	熟练掌握四合院的分类、宅门及影壁制式、四合院各部分名称					
学习目标	1. 熟练掌握四合院的分类； 2. 熟练掌握四合院宅门的制式； 3. 熟练掌握四合院影壁的制式； 4. 熟练掌握四合院各组成部分名称					
任务描述	熟练掌握四合院的分类、宅门及影壁制式、四合院各部分名称					
学时安排	情境导入	计划	决策	实施	检查	评价
时长	10分钟	10分钟	10分钟	30分钟	15分钟	15分钟
对学生的要求	1. 课前做好项目预习； 2. 做好课中总结与评价； 3. 通过四合院的学习内容谈谈建筑语言中的人文主义精神及中华民族传统文化					
参考资料	1. 微课视频； 2. PPT； 3. 课程相关视频资料					

项目前学习任务单

学习领域	四合院		用时	
达成目标	1. 观看微课视频，可以简单整理出课程内容思维导图； 2. 根据课程内容，感受中国古建筑文化的博大精深及大国工匠精神			
学习方法	看：观看微课视频及相关课程资料； 思：认真思考任务单中的问题； 记：将心中的疑惑记下来			
课堂学习形式预告	课堂外		课堂内	
	1. 自主学习，搜集资料； 2. 整理收获，提出问题		1. 展示交流； 2. 协作探究； 3. 启发式教学； 4. 项目驱动； 5. 实习作业法	
本次学习任务	任务1：熟练掌握四合院的分类； 任务2：熟练掌握四合院宅门的制式； 任务3：熟练掌握四合院影壁的制式； 任务4：熟练掌握四合院各组成部分名称			
疑点与问题	本次学习中还有哪些不明白的地方？请依次记录下来。			

总结评价	1. 你认为自己预习得怎么样？
	  A. 马马虎虎，只是为了完成任务
	  B. 态度认真，收获很多
	2. 你对自己本次自主学习的评价：☆☆☆☆☆			

资讯单

学习领域	四合院		
学习情境	学习四合院的相关知识点		
学习任务	熟练掌握四合院的分类、宅门及影壁制式、四合院各部分名称及功能	学时	10 分钟
资讯方式	1. 教师讲解； 2. 互联网查询； 3. 同学交流		
资讯问题	1. 四合院的分类有哪些？ 2. 四合院的宅门制式有哪些？ 3. 四合院的影壁制式包括哪些？ 4. 四合院的各组成部分的名称及功能是什么？ 5. 通过古建筑四合院的学习内容谈谈建筑语言中的人文主义精神及中华民族传统文化		
资讯引导	1. 微课视频； 2. PPT； 3. 课程相关视频资料		

计划单

学习领域	四合院		
学习情境	学习四合院的相关知识点		
学习任务	熟练掌握四合院的分类、宅门及影壁制式、四合院各部分名称及功能	学时	10 分钟
典型工作过程描述	教学情境导入—四合院的分类—宅门的形制—影壁的形制—"三进式"四合院的各组成部分名称及功能—民族文化精神解读		
计划制订的方式	同学间分组讨论		

序号	工作步骤	注意事项
1	查看资料文件	
2	查询资料	
3	设计课程内容思维导图	

计划评价	班级		第___组	组长签字	
	教师签字		日期		
	评语：				

决策单

学习领域	四合院		
学习情境	学习四合院的相关知识点		
学习任务	熟练掌握四合院的分类、宅门及影壁制式、四合院各部分名称及功能	学时	10分钟
典型工作过程描述	教学情境导入—四合院的分类—宅门的形制—影壁的形制—"三进式"四合院的各组成部分名称及功能—民族文化精神解读		

学生为主体的课堂决策

序号	课程目标	课程内容	教学活动	教学资料	综合评价
1					
2					
3					
4					
5					
6					
7					
8					
9					
10					

决策评价	班级		第___组	组长签字	
	教师签字		日期		
	评语：				

实施单

学习领域	四合院		
学习情境	学习四合院的相关知识点		
学习任务	熟练掌握四合院的分类、宅门及影壁制式、四合院各部分名称及功能	学时	30 分钟
典型工作过程描述	教学情境导入—四合院的分类—宅门的形制—影壁的形制—"三进式"四合院的各组成部分名称及功能—民族文化精神解读		

序号	实施步骤	注意事项
1	教学情境导入	播放先导片
2	四合院的分类	教学重点与难点
3	宅门及影壁的制式	教学重点与难点
4	四合院各组成部分名称	教学重点与难点
5	结合课程思政	引入中华民族传统文化精神

实施说明：
1. 播放先导片；
2. 借助微课视频与 PPT 进行深入了解；
3. 通过了解四合院的形制引入中华民族传统文化精神。

实施评价	班级		第＿＿组	组长签字	
	教师签字		日期		
	评语：				

检查单

学习领域	四合院		
学习情境	学习四合院的相关知识点		
学习任务	熟练掌握四合院的分类、宅门及影壁制式、四合院各部分名称及功能	学时	15 分钟
典型工作过程描述	教学情境导入—四合院的分类—宅门的形制—影壁的形制—"三进式"四合院的各组成部分名称及功能—民族文化精神解读		

序号	检查项目	检查标准	学生自查	教师检查
1	资讯环节	获取相关信息情况		
2	计划环节	设计课程思维导图		
3	实施环节	绘制课程思维导图		
4	检查环节	各个环节逐一检查		

检查评价	班级		第＿＿组	组长签字	
	教师签字		日期		
	评语：				

评价单

学习领域	四合院			
学习情境	学习四合院的相关知识点			
学习任务	熟练掌握四合院的分类、宅门及影壁制式、四合院各部分名称及功能		学时	15分钟
典型工作过程描述	教学情境导入—四合院的分类—宅门的形制—影壁的形制—"三进式"四合院的各组成部分名称及功能—民族文化精神解读			
评价项目	评价子项目	学生自评	组内评价	教师评价
资讯环节	1. 听取教师讲解； 2. 互联网查询情况； 3. 同学交流情况			
计划环节	1. 查询资料情况； 2. 设计思维导图			
实施环节	1. 学习态度； 2. 课程内容熟练程度； 3. 课堂表现； 4. 作业完成情况及质量			
最终结果	综合情况			
评价	班级		第___组	组长签字	
	教师签字		日期		
	评语：				

13.2 理论指导

13.2.1 四合院的分类

四合院

四合院又称为四合房,是中国的一种传统合院式建筑,其格局为一个庭院四面建有房屋,从四面将庭院合围在中间。四合院是按照院落数量进行分类的,可以分为一进院落、二进院落、三进院落、四进院落、五进院落。

呈"口"字形的称为一进院落,整个院落是"口"字的形状;呈"日"字形的称为二进院落;呈"目"字形的称为三进院落;四进院落的整个院子被分为四段;现存的五进院落很稀有。完整的四合院为三进院落,是需要重点了解的四合院形制。

四合院由前向后依次是宅门、宅门旁边的倒座房,之后是影壁、影壁后面的垂花门,接着是连着垂花门的游廊,两边是厢房,然后是正房,正房两侧是耳房以及最里面的后罩房(图13-1)。

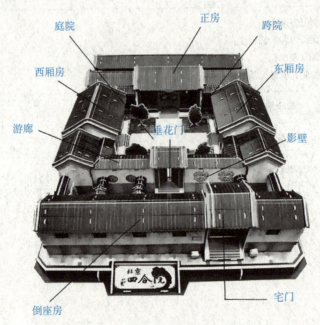

图 13-1 四合院"三进院落"布局

13.2.2 宅门的制式

宅门是指院子临街的大门,是旧社会主人地位的表象特征,这里面的大门一般分为王府大门、广亮大门、金柱大门、蛮子门及如意门等。

1. 王府大门

王府大门为四合院最高级别的宅门,是皇家宗室专属用门。最常见的有五间三启门和三间一启门两种规格（图13-2）。

2. 广亮大门

广亮大门是四合院中最具代表性的大门，是一定品级的官员才能使用的宅门。广亮大门的台基较高，其构架的主体为前、后檐柱和中柱（图13-3）。

图13-2　王府大门

图13-3　广亮大门

3. 金柱大门

金柱大门是具有一定品级的官宦之家才能使用的宅门。金柱大门设置在中柱与前檐柱之间的金柱上，相对于广亮大门宽阔的门前空间，金柱大门的门前空间较小（图13-4）。

4. 蛮子门

蛮子门将门扉直接安置在前檐柱之上，相传是为了避免贼人在门前停留。"蛮子"在古汉语中意为粗野、野蛮，蛮子门之名也由此而来（图13-5）。

5. 如意门

如意门是四合院普遍使用的宅门，它是在前檐柱之间砌墙，在如意门门口设有两个门簪，多刻有"如意"二字，这也是如意门名称的由来（图13-6）。

图13-4　金柱大门

图13-5　蛮子门

图13-6　如意门

13.2.3　影壁的制式

影壁也称作照壁，是中国传统建筑中用于遮挡视线、增加住宅气势的墙壁，古代人很重视保护隐私。

1. 一字影壁

一字影壁俯视平面为单一的细长条，有的一字影壁因太长而显得单调，则将其横向分为三段，中间长两头短，形成"一主二从"的形式（图13-7）。

2. 八字影壁

八字影壁在一字影壁的基础上，将"二从"向内折，形成八字围合状（图13-8）。

3. 座山影壁

座山影壁是在厢房的山墙上直接砌出小墙帽并做出影壁形状，使影壁与山墙连为一体（图13-9）。

4. 撇山影壁

撇山影壁是"反八字"影壁，做这种影壁时，大门要向里退2～4m，在门前形成一个小空间，以作为进出大门的缓冲之地，这也是撇山影壁与八字影壁最明显的区别（图13-10）。

图13-7　一字影壁

图13-8　八字影壁

图13-9　座山影壁

图13-10　撇山影壁

13.2.4　四合院各组成部分的名称及功能

1. 宅门

宅门是指院子临街的大门，是旧社会主人地位的一个表象特征。

2. 倒座房

与宅门相连接的是倒座房，倒座房紧邻四合院宅门，它是坐南朝北与正房相对的房子，又称南房，通常为外客厅、账房、门房、客房，因此它的用途多为供客人或仆人居住（图13-11）。

3. 影壁

影壁也称作照壁，是中国传统建筑中用于遮挡视线、增加住宅气势的墙壁。

4. 垂花门

经过影壁之后就会看见垂花门，垂花门是内宅与外宅的分界线和唯一通道，进入垂花门就是院落内部，外人一般不得随便出入。古人常说的"大门不出，二门不迈"中的"二门"指的就是垂花门。垂花门上的檐柱是不落地的，整座建筑占天不占地，这是垂花门的特色之一（图 13-12）。

图 13-11　倒座房

图 13-12　垂花门

5. 抄手游廊

抄手游廊也叫作游廊，抄手游廊是四合院中走廊的常见形式，与垂花门衔接，环绕垂花门、厢房和正房。它的设计便于人们在雨雪天中行走（图 13-13）。

6. 厢房

厢房为院子东、西两侧的房子，分为东、西厢房，通常长子住东厢房，次子住西厢房（图 13-14）。

7. 正房

正房是四合院的核心建筑，是一家之主的居所，一般为三间。正房的明间，也就是中间的一间称为堂屋，也称为中堂；三开间的正房堂屋两侧一般为卧室和书房。正房的特点是冬天太阳能够照进屋里，冬暖夏凉（图 13-15）。

图 13-13　抄手游廊

图 13-14　厢房

图 13-15　正房

8. 耳房

正房的两侧还各有一间或两间进深、高度都偏小的房间，如同挂在正房两侧的两只耳朵，叫作耳房。耳房可以作为仓库与厨房（图 13-16）。

9. 后罩房

后罩房在正房后面，有一层的，也有两层的，均坐北朝南，与正房后山墙之间形成一个院子，一般称为后院。后院供宅主人的内眷或老人居住（图 13-17）。

图 13-16　耳房

图 13-17　后罩房

13.3　中华民族传统文化精神解读

　　四合院一般是一户一住，但也有多户（多为贫困人家）合住一座四合院的情况，这种四合院称为大杂院。大杂院的温馨是许多老北京居民无法忘记的。四合院虽为居住建筑，却蕴含着深刻的文化内涵，是中华传统文化的载体。四合院的装修、雕饰、彩绘处处体现着民俗民风和传统文化，表现了一定历史条件下人们对幸福、美好、富裕、吉祥的追求。

13.4　项目创新

　　绘制一张三进四合院的动线图，并标注每一部分的中、英文名称。

项目 14　大宅门——乔家大院

14.1　项目表单

学习任务单

学习领域	乔家大院					
学习情境	学习乔家大院的相关知识点					
学习任务	熟练掌握乔家大院的基本建筑形制、六院布局及特色，三宝院中的 3 个宝物名称			学时	2 学时（90 分钟）	
教学目标	使学生充分了解乔家大院的基本建筑形制、六院布局及特色，三宝院中的 3 个宝物名称					
技能目标	通过乔家大院进一步了解中国古建筑结构的精髓及中国优良的传统文化					
知识目标	熟练掌握乔家大院的基本建筑形制、六院布局及特色，三宝院中的 3 个宝物名称					
学习目标	1. 乔家大院的基本建筑形制； 2. 乔家大院中的六院布局及特色； 3. 三宝院中的 3 个宝物的名称及特点					
任务描述	熟练掌握乔家大院的基本建筑形制、六院布局及特色，三宝院中的 3 个宝物名称					
学时安排	情境导入	计划	决策	实施	检查	评价
时长	10 分钟	10 分钟	10 分钟	30 分钟	15 分钟	15 分钟
对学生的要求	1. 课前做好项目预习； 2. 做好课中总结与评价； 3. 通过乔家大院的学习内容谈谈建筑语言中的人文主义精神及中华民族传统文化					
参考资料	1. 微课视频； 2. PPT； 3. 课程相关视频资料					

项目前学习任务单

学习领域	乔家大院		用时	
达成目标	1. 观看微课视频，可以简单整理出课程内容思维导图； 2. 根据课程内容，感受中国古建筑文化的博大精深及大国工匠精神			
学习方法	看：观看微课视频及相关课程资料； 思：认真思考任务单中的问题； 记：将心中的疑惑记下来			
课堂学习形式预告	课堂外		课堂内	
	1. 自主学习，搜集资料； 2. 整理收获，提出问题		1. 展示交流； 2. 协作探究； 3. 启发式教学； 4. 项目驱动； 5. 实习作业法	
本次学习任务	任务1：乔家大院的基本建筑形制； 任务2：乔家大院中的六院布局及特色； 任务3：三宝院中的3个宝物名称及特点			
疑点与问题	本次学习中还有哪些不明白的地方？请依次记录下来。			
总结评价	1. 你认为自己预习得怎么样？ 　A. 马马虎虎，只是为了完成任务 　B. 态度认真，收获很多 2. 你对自己本次自主学习的评价：☆☆☆☆☆			

项目 14　大宅门——乔家大院

资讯单

学习领域	乔家大院		
学习情境	学习乔家大院的相关知识点		
学习任务	熟练掌握乔家大院的基本建筑形制、六院布局及特色，三宝院中的 3 个宝物名称	学时	10 分钟
资讯方式	1. 教师讲解； 2. 互联网查询； 3. 同学交流		
资讯问题	1. 乔家大院的基本建筑形制是什么？ 2. 乔家大院中的六院布局及特色是什么？ 3. 三宝院中的 3 个宝物分别是什么？ 4. 通过古建筑乔家大院的学习内容谈谈建筑语言中的人文主义精神及中华民族传统文化		
资讯引导	1. 微课视频； 2. PPT； 3. 课程相关视频资料		

计划单

学习领域	乔家大院		
学习情境	学习乔家大院的相关知识点		
学习任务	熟练掌握乔家大院的基本建筑形制、六院布局及特色，三宝院中的 3 个宝物名称	学时	10 分钟
典型工作过程描述	教学情境导入—乔家大院的基本建筑形制—乔家发迹史—六院布局及特色—民族文化精神解读		
计划制订的方式	同学间分组讨论		
序号	工作步骤	注意事项	
1	查看资料文件		
2	查询资料		
3	设计课程内容思维导图		
计划评价	班级　　　　第　　组　　组长签字 教师签字　　　　　　日期 评语：		

决策单

学习领域	乔家大院		
学习情境	学习乔家大院的相关知识点		
学习任务	熟练掌握乔家大院的基本建筑形制、六院布局及特色，三宝院中的3个宝物名称	学时	10分钟
典型工作过程描述	教学情境导入—乔家大院的基本建筑形制—乔家发迹史—六院布局及特色—民族文化精神解读		

学生为主体的课堂决策

序号	课程目标	课程内容	教学活动	教学资料	综合评价
1					
2					
3					
4					
5					
6					
7					
8					
9					
10					

决策评价	班级		第___组	组长签字	
	教师签字		日期		
	评语：				

实施单

学习领域	乔家大院			
学习情境	学习乔家大院的相关知识点			
学习任务	熟练掌握乔家大院的基本建筑形制、六院布局及特色,三宝院中的3个宝物名称		学时	30分钟
典型工作过程描述	教学情境导入—乔家大院的基本建筑形制—乔家发迹史—六院布局及特色—民族文化精神解读			
序号	实施步骤		注意事项	
1	教学情境导入		播放先导片	
2	乔家大院的基本建筑形制		教学重点与难点	
3	六院布局及特色		教学重点与难点	
4	三宝院中的3个宝物名称及特色		教学重点与难点	
5	结合课程思政		引入中华民族传统文化精神	

实施说明：
1. 播放先导片；
2. 借助微课视频与PPT进行深入了解；
3. 通过了解乔家大院引入中华民族传统文化精神

实施评价	班级		第___组	组长签字	
	教师签字		日期		
	评语：				

检查单

学习领域	乔家大院			
学习情境	学习乔家大院的相关知识点			
学习任务	熟练掌握乔家大院的基本建筑形制、六院布局及特色,三宝院中的3个宝物名称		学时	15分钟
典型工作过程描述	教学情境导入—乔家大院的基本建筑形制—乔家发迹史—六院布局及特色—民族文化精神解读			
序号	检查项目	检查标准	学生自查	教师检查
1	资讯环节	获取相关信息情况		
2	计划环节	设计课程思维导图		
3	实施环节	绘制课程思维导图		
4	检查环节	各个环节逐一检查		

检查评价	班级		第___组	组长签字	
	教师签字		日期		
	评语：				

评价单

学习领域	乔家大院			
学习情境	学习乔家大院的相关知识点			
学习任务	熟练掌握乔家大院的基本建筑形制、六院布局及特色,三宝院中的3个宝物名称		学时	15分钟
典型工作过程描述	教学情境导入—乔家大院的基本建筑形制—乔家发迹史—六院布局及特色—民族文化精神解读			
评价项目	评价子项目	学生自评	组内评价	教师评价
资讯环节	1. 听取教师讲解; 2. 互联网查询情况; 3. 同学交流情况			
计划环节	1. 查询资料情况; 2. 设计思维导图			
实施环节	1. 学习态度; 2. 课程内容熟练程度; 3. 课堂表现; 4. 作业完成情况及质量			
最终结果	综合情况			
评价	班级		第___组	组长签字	
	教师签字		日期		
	评语:				

14.2 理论指导

乔家大院

14.2.1 乔家大院的基本建筑制式

乔家大院又名在中堂，位于山西省祁县乔家堡村，始建于1756年，整个院落呈双"喜"字形，分为6个大院，内套20个小院，共有313间房屋，建筑面积为4 175 m²。乔家大院三面临街，四周是高达十余米的全封闭青砖墙，是一座具有北方传统民居建筑风格的古宅。

14.2.2 乔家大院的六院布局及建筑特色

1. 第一院

第一院也称为老院，由乔致庸的父亲乔全美所建。

从第一院进入大门之后有一个照壁，叫作福德祠，这个照壁印证了"门迎百寿，院纳福德"，它的用途一是装饰，二是镇宅辟邪。第一院的正门上挂着一个匾额，匾额上刻着"毋不敬"，毋不敬也就是勿不敬，出自朱子《礼记》第一篇《曲礼》中的第一句话，可以理解为：天地之间有神灵，不敬的态度是不可取的（图14-1）。

2. 第二院

第二院展现的就是乔家的发迹史，这里的院落即呈双"喜"字构建，门前匾额上雕刻着"会芳"，寓意贤德之士汇聚一堂。在匾额下面的楹联上刻着"宽宏坦荡福臻家常裕，温厚和平荣久后必昌"，它的意思是处事心地宽宏坦荡就会福至运达，家族时常富裕；待人性情温厚平和，才能荣耀久长，后辈必定昌盛。这里充分体现了乔家严格的家规家训和晋商精神（图14-2）。

图14-1 乔家大院第一院

图14-2 乔家大院第二院

3. 第三院

第三院也称为三宝院，其中隐藏了3个镇宅之宝。

（1）万人球。万人球是一个挂在天花板上的水银玻璃球，据说是乔家从美国购来的。这个玻璃球会按照一定比例缩小屋中人的影像并映在球上。当年乔家会客谈生意时将其挂在房上用以监视屋中人。万人球也被称为世界上第一个监视探头，为当时极罕见之物

（图14-3）。

（2）犀牛望月镜。犀牛望月镜是一面直径为1 m左右的镜子，镜架镜框为木质最坚硬的珍稀树木——铁力木所制，高2 m，重1 t，整个宝物由3部分组成：上面是镜子，代表圆圆的月亮；中间部分是祥云；底座是犀牛，这里"犀牛"的谐音是"喜牛"，是大吉大利、喜从天降的吉祥物。犀牛望月镜被国家列为一级文物（图14-4）。

（3）九龙灯。九龙灯由清代红木制成。当时慈禧太后在八国联军侵华而逃向西安时，乔家捐赠30万两银子以解慈禧太后的燃眉之急，之后慈禧太后为感谢其忠诚而赏赐其两盏九龙灯，至今全国未发现第三盏（图14-5）。

图14-3　万人球　　　　　图14-4　犀牛望月镜　　　　　图14-5　九龙灯

4. 第四院

第四院是商俗院，也是新院。它由乔致庸之孙乔映霞所建，因为乔映霞早年有留学经历，所以这个院子在建筑风格上处处体现出他留学所带来的先进技术。最具特色的就是东厢房屋檐下方的雕刻，内容为一列火车正呼啸着在桥上行驶，这是乔映霞凭借自己记忆所画，因为当年中国还没有火车（图14-6）。

门楣上方的砖雕，分别刻有琴棋书画，寓意是乔家人希望自己的子孙多才多艺，可以琴棋书画样样精通（图14-7）。

图14-6　第四院屋檐下方的雕刻　　　　图14-7　第四院门楣上方的砖雕

5. 第五院

第五院也叫作在中堂，由乔致庸所建。门楣上方有一块匾额，刻着"在中堂"。它取意为"不偏不倚，执用之中"，体现了主人的中庸思想；门楣上也雕刻了福禄寿三星，两边有八骏的雕刻，寓意是希望后代子孙都有所作为。第五院也展现了主人的人生礼仪。

门楣下面的楹联上刻着"传家有道唯存厚，处世无奇但率真"，意思是唯有积存厚德才是传承家业的真道，处世立身没有奇特的方法，只有依循直率真诚的本性。这充分说明了乔家重厚道，讲真诚，不欺不诈，不瞒不骗，以义取利。诚信聚人气，重德汇财源，简洁的联语点明了主人传家与处世之道（图 14-8）。

6. 第六院

乔家主人本来要在第六院建一座"书房院"，当时已经买回了全套假山。但是 1937 年"七七事变"爆发，这里只留下一部分花园，改成了"花园院"，后留下了未完成的花园（图 14-9）。

书房用于培养乔家子弟，希望他们发奋读书，增长知识，成为优秀人才。书房院与祠堂紧邻，可以随时随地激励乔家子孙后代学业进步。

图 14-8　第五院（在中堂）

图 14-9　第六院（花园院）

14.3　中华民族传统文化精神解读

乔家大院充分体现了中国晋商文化的精神，在这里看不到传统文化与商人文化的对抗。乔致庸治理下的乔家，以"经商者先立人，唯忠义者是商本"为标准，进一步补充了传统文化里关于修身、齐家、治国的理论，并且把自我修养、自我完善和家族精神、开拓进取结合起来。如此，众多美好的品格就纷至沓来——勤俭、诚信、包容、克己，而它们也一路伴随着这个家族走向辉煌，纵使历史的发展曲折，天灾人祸频仍，但这个家族始终保持昂扬的斗志，在这美丽富饶的土地上聚散流沙，生生不息。

14.4　项目创新

谈一谈你的感想，包括对晋商文化精神的理解，可以用短视频的方式记录你的感想，或者写 500 字的文字记录。

项目 15　徽州民居

15.1　项目表单

学习任务单

学习领域	徽州民居					
学习情境	学习徽州民居的相关知识点					
学习任务	熟练掌握徽州民居的建筑制式、与四合院形制的区别及自身建筑特色与精神文化内涵				学时	2学时（90分钟）
教学目标	使学生充分了解徽州民居的建筑制式、与四合院形制的区别及自身建筑特色与精神文化内涵					
技能目标	通过徽州民居进一步了解中国古建筑结构的精髓及中国优良的传统文化					
知识目标	熟练掌握徽州民居的建筑制式、与四合院形制的区别及自身建筑特色					
学习目标	1. 熟练掌握徽州民居的建筑制式； 2. 熟练掌握徽州民居与四合院在建筑制式上的区别； 3. 熟练掌握徽州民居的建筑特色					
任务描述	熟练掌握徽州民居的建筑制式、与四合院形制的区别及自身建筑特色与精神文化内涵					
学时安排	情境导入	计划	决策	实施	检查	评价
时长	10分钟	10分钟	10分钟	30分钟	15分钟	15分钟
对学生的要求	1. 课前做好项目预习； 2. 做好课中总结与评价； 3. 通过徽州民居的学习内容谈谈建筑语言中的人文主义精神及中华民族传统文化					
参考资料	1. 微课视频； 2. PPT； 3. 课程相关视频资料					

项目前学习任务单

学习领域	徽州民居		用时	
达成目标	1. 观看微课视频，可以简单整理出课程内容思维导图； 2. 根据课程内容，感受中国古建筑文化的博大精深及大国工匠精神			
学习方法	看：观看微课视频及相关课程资料； 思：认真思考任务单中的问题； 记：将心中的疑惑记下来			
课堂学习形式预告	课堂外		课堂内	
	1. 自主学习，搜集资料； 2. 整理收获，提出问题		1. 展示交流； 2. 协作探究； 3. 启发式教学； 4. 项目驱动； 5. 实习作业法	
本次学习任务	任务 1：徽州民居的基本建筑制式； 任务 2：徽州民居与四合院在建筑制式上的区别； 任务 3：徽州民居的建筑特色			
疑点与问题	本次学习中还有哪些不明白的地方？请依次记录下来。 			

总结评价	1. 你认为自己预习得怎么样？
	  A. 马马虎虎，只是为了完成任务
	  B. 态度认真，收获很多
	2. 你对自己本次自主学习的评价：☆☆☆☆☆			

资讯单

学习领域	徽州民居		
学习情境	学习徽州民居的相关知识点		
学习任务	熟练掌握徽州民居的建筑制式、与四合院形制的区别及自身建筑特色与精神文化内涵	学时	10分钟
资讯方式	1. 教师讲解； 2. 互联网查询； 3. 同学交流		
资讯问题	1. 徽州民居建筑布局分类有哪些？ 2. 徽州民居与四合院在建筑制式上有什么区别？ 3. 徽州民居的建筑特色有哪些？ 4. 通过古建筑徽州民居的学习内容谈谈建筑语言中的人文主义精神及中华民族传统文化		
资讯引导	1. 微课视频； 2. PPT； 3. 课程相关视频资料		

计划单

学习领域	徽州民居		
学习情境	学习徽州民居的相关知识点		
学习任务	熟练掌握徽州民居的建筑制式、与四合院形制的区别及自身建筑特色与精神文化内涵	学时	10分钟
典型工作过程描述	教学情境导入—徽州民居的基本建筑制式—徽州民居与四合院形制的区别—徽州民居的建筑特色—民族文化精神解读		
计划制订的方式	同学间分组讨论		

序号	工作步骤	注意事项
1	查看资料文件	
2	查询资料	
3	设计课程内容思维导图	

	班级		第___组	组长签字
计划评价	教师签字		日期	
	评语：			

决策单

学习领域	徽州民居		
学习情境	学习徽州民居的相关知识点		
学习任务	熟练掌握徽州民居的建筑制式、与四合院形制的区别及自身建筑特色与精神文化内涵	学时	10分钟
典型工作过程描述	教学情境导入—徽州民居的基本建筑制式—徽州民居与四合院形制的区别—徽州民居的建筑特色—民族文化精神解读		

学生为主体的课堂决策

序号	课程目标	课程内容	教学活动	教学资料	综合评价
1					
2					
3					
4					
5					
6					
7					
8					
9					
10					

决策评价	班级		第___组	组长签字	
	教师签字		日期		
	评语:				

实施单

学习领域	徽州民居		
学习情境	学习徽州民居的相关知识点		
学习任务	熟练掌握徽州民居的建筑制式、与四合院形制的区别及自身建筑特色与精神文化内涵	学时	30分钟
典型工作过程描述	教学情境导入—徽州民居的基本建筑制式—徽州民居与四合院形制的区别—徽州民居的建筑特色—民族文化精神解读		

序号	实施步骤	注意事项
1	教学情境导入	播放先导片
2	徽州民居的建筑形制	教学重点与难点
3	与四合院形制的区别	教学重点与难点
4	徽州民居的建筑特色	教学重点与难点
5	结合课程思政	引入中华民族传统文化精神

实施说明:
1. 播放先导片;
2. 借助微课视频与PPT进行深入了解;
3. 通过了解徽州民居引入中华民族传统文化精神

实施评价	班级		第___组	组长签字	
	教师签字		日期		
	评语:				

检查单

学习领域	徽州民居		
学习情境	学习徽州民居的相关知识点		
学习任务	熟练掌握徽州民居的建筑制式、与四合院形制的区别及自身建筑特色与精神文化内涵	学时	15分钟
典型工作过程描述	教学情境导入—徽州民居的基本建筑制式—徽州民居与四合院形制的区别—徽州民居的建筑特色—民族文化精神解读		

序号	检查项目	检查标准	学生自查	教师检查
1	资讯环节	获取相关信息情况		
2	计划环节	设计课程思维导图		
3	实施环节	绘制课程思维导图		
4	检查环节	各个环节逐一检查		

检查评价	班级		第___组	组长签字	
	教师签字		日期		
	评语:				

评价单

学习领域	徽州民居			
学习情境	学习徽州民居的相关知识点			
学习任务	熟练掌握徽州民居的建筑制式、与四合院形制的区别及自身建筑特色与精神文化内涵		学时	15 分钟
典型工作过程描述	教学情境导入—徽州民居的基本建筑制式—徽州民居与四合院形制的区别—徽州民居的建筑特色—民族文化精神解读			
评价项目	评价子项目	学生自评	组内评价	教师评价
资讯环节	1. 听取教师讲解； 2. 互联网查询情况； 3. 同学交流情况			
计划环节	1. 查询资料情况； 2. 设计思维导图			
实施环节	1. 学习态度； 2. 课程内容熟练程度； 3. 课堂表现； 4. 作业完成情况及质量			
最终结果	综合情况			
评价	班级		第___组	组长签字
	教师签字		日期	
	评语：			

15.2 理论指导

15.2.1 徽州民居基本建筑形制

徽州民居

徽州民居，也称为徽派民居，指徽州地区具有徽州传统风格的民居，是中国传统民居建筑的一个重要流派。因为徽州地区被誉为天然公园，所以徽州民居的建筑制式与自然风光浑然天成，使实用性与艺术性得到完美统一。

徽州民居的基本建筑形制分为"凹字形"布局、"回字形"布局、"H形"布局及"日字形"布局。

1. "凹字形"布局

"凹字形"，即三合院式，又称为三间式，是徽州民居中最为经济的一种，多为一进的两层住宅，天井位于中央，楼下明间作客厅，楼上明间为供奉祖先神位的祖堂。两侧厢房可作卧室（图 15-1）。

2. "回字形"布局

"回字形"，又称为"口字形"、四合院式，为三间两进，门厅楼上明间为正间，后进厅堂楼上明间一般作为祖堂，两侧厢房是卧室（图 15-2）。

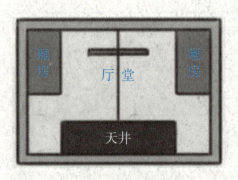

图 15-1 "凹字形"布局

图 15-2 "回字形"布局

3. "H形"布局

"H形"，即前后两个三合院背向型。其为三间两进，前、后各有一天井。这种住宅没有门厅，正堂往往分为两个厅堂，两个厅堂合用一个屋脊，俗称"一脊翻两堂"（图 15-3）。

4. "日字形"布局

"日字形"，为三间三进，共有两个天井，各进之间两边有廊庑相连。廊庑指的是廊屋，祖堂一般设在最后一进厅堂（图15-4）。

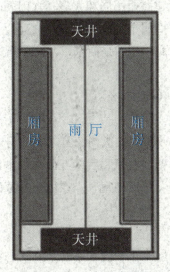
图15-3 "H形"布局

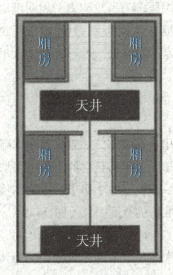
图15-4 "日字形"布局

15.2.2 徽州民居与四合院布局的区别

徽州民居与四合院在建筑制式上的区别如下。以"三进式"布局为例，徽州民居不是屋套屋的布局，它与四合院的最大区别在于天井，四合院是没有天井的，而每一个徽州民居形制中都有天井。

15.2.3 徽州民居建筑的特点

1. 屋套屋

徽州民居的建筑制式都是一屋套一屋的，形成屋套屋的建筑布局（图15-5）。

2. 天井

天井的形状一般为长方形。天井的作用是采光、通风，因为徽州民居四周的墙壁很高，窗口又比较狭窄，所以人们会利用天井来采光、通风。它还有排水功能，因此有"四水归堂"的吉祥寓意（图15-6）。

3. 马头墙

马头墙，又称为风火墙、防火墙、封火墙。马头墙特指山墙的墙顶部分，因形状酷似马头，故称"马头墙"。江南传统民居建筑的墙体之所以采取这种形式，主要是因为在聚族而居的村落中，民居建筑密度较大，火灾发生时，火势容易顺房蔓延，马头墙可以有效阻止火势蔓延（图15-7）。

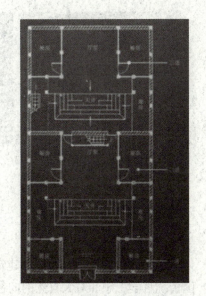
图15-5 "屋套屋"建筑布局特色

4. 朝北居

一般徽州民居多为坐北朝南，但明清时期所建徽州民居却大多是大门朝北。原来古徽州人的居住习惯有许多禁忌。这里据五行说法：商属金，南方属火，火克金，不吉利；征属火，北方属水，水克火，也不吉利。这里的"商"指的是商家，"征"指的是兵家。

5. 重檐

重檐又称为短檐，位于屋檐下方、门窗上方（图15-8）。

6. 满顶床

满顶床采用全木结构，是兼具徽州木雕、徽州漆器、徽州家具之美的房式老床。

所谓"满顶"，指的是它的上顶、下底、左壁、右壁和后壁五面都以木板满封、密不透风。它就像一幢独立的小房子，成为"房中房""室中室"。在徽州人家的厢房中，它的尺寸正好与厢房的宽度一致（图15-9）。

7. 压画桌

压画桌为徽州民居中的传统陈设。徽州民居厅堂正中壁上多挂中堂画、对联，均装裱成卷轴悬挂。在卷轴之下设长条桌，桌面上放置两个马鞍形的画脚，卷轴向下展开至长条桌，搁入画脚的"马鞍"内，画幅即平整稳固，此长条桌即压画桌（图15-10）。

图 15-6　天井

图 15-7　马头墙

图 15-8　重檐

图 15-9　满顶床

图 15-10　压画桌

15.3　中华民族传统文化精神解读

徽州民居的选址、布局和建筑形态，都以周易风水理论为指导，体现了天人合一的中国传统哲学思想和对大自然的向往与尊重。那些典雅的明清民居建筑群与大自然紧密相融，创造出一个既合乎科学，又富有情趣的生活居住环境，是中国传统民居的精髓。村落独特的水系是实用与美学相结合的水利工程典范，深刻体现了人类利用自然、改造自然的卓越智慧。其"布局之工，结构之巧，装饰之美，营造之精，文化内涵之深"，为国内古民居建筑群所罕见。

15.4　项目创新

回顾本节课中的徽州民居建筑特色，与之前所学的北方民居四合院做对比，包括建筑形制、风格与特点等，并写一份报告。

项目 16　福建土楼

16.1　项目表单

学习任务单

学习领域	福建土楼					
学习情境	学习福建土楼的相关知识点					
学习任务	熟练掌握福建土楼群的名称、承启楼的建筑制式、田螺坑土楼群的组成部分及土楼的作用				学时	2学时（90分钟）
教学目标	使学生充分了解福建土楼群的名称、承启楼的建筑制式、田螺坑土楼群的组成部分及土楼的作用					
技能目标	通过福建土楼进一步了解中国古建筑结构的精髓及中国优良的传统文化					
知识目标	熟练掌握福建土楼群的名称、承启楼的建筑制式、田螺坑土楼群的组成部分及土楼的作用					
学习目标	1. 承启楼的建筑制式与特点分别是什么？ 2. 土楼的建筑材料及作用是什么？ 3. 田螺坑土楼群分别由哪几座土楼组合而成？					
任务描述	熟练掌握福建土楼群的名称、承启楼的建筑制式、田螺坑土楼群的组成部分及土楼的作用					
学时安排	情境导入	计划	决策	实施	检查	评价
时长	10分钟	10分钟	10分钟	30分钟	15分钟	15分钟
对学生的要求	1. 课前做好项目预习； 2. 做好课中总结与评价； 3. 通过福建土楼的学习内容谈谈建筑语言中的人文主义精神及中华民族传统文化					
参考资料	1. 微课视频； 2. PPT； 3. 课程相关视频资料					

项目 16　福建土楼

项目前学习任务单

学习领域	福建土楼		用时	
达成目标	1. 观看微课视频，可以简单整理出课程内容思维导图； 2. 根据课程内容，感受中国古建筑文化的博大精深及大国工匠精神			
学习方法	看：观看微课视频及相关课程资料； 思：认真思考任务单中的问题； 记：将心中的疑惑记下来			
课堂学习形式预告	课堂外		课堂内	
	1. 自主学习，搜集资料； 2. 整理收获，提出问题		1. 展示交流； 2. 协作探究； 3. 启发式教学； 4. 项目驱动； 5. 实习作业法	
本次学习任务	任务 1：承启楼的建筑制式与特点分别是什么？ 任务 2：土楼的建筑材料及作用是什么？ 任务 3：田螺坑土楼群分别由哪几座土楼组合而成？			
疑点与问题	本次学习中还有哪些不明白的地方？请依次记录下来。			
总结评价	1. 你认为自己预习得怎么样？ 　A. 马马虎虎，只是为了完成任务 　B. 态度认真，收获很多 2. 你对自己本次自主学习的评价：☆☆☆☆☆			

资 讯 单

学习领域	福建土楼		
学习情境	学习福建土楼的相关知识点		
学习任务	熟练掌握福建土楼群的名称、承启楼的建筑制式、田螺坑土楼群的组成部分及土楼的作用	学时	10 分钟
资讯方式	1. 教师讲解； 2. 互联网查询； 3. 同学交流		
资讯问题	1. 承启楼的建筑制式与特点分别是什么？ 2. 土楼的建筑材料及作用是什么？ 3. 田螺坑土楼群分别由哪几座土楼组合而成？ 4. 通过古建筑福建土楼的学习内容谈谈建筑语言中的人文主义精神及中华民族传统文化		
资讯引导	1. 微课视频； 2. PPT； 3. 课程相关视频资料		

计 划 单

学习领域	福建土楼		
学习情境	学习福建土楼的相关知识点		
学习任务	熟练掌握福建土楼群的名称、承启楼的建筑制式、田螺坑土楼群的组成部分及土楼的作用	学时	10 分钟
典型工作过程描述	教学情境导入—福建土楼群的分布及名称—承启楼的建筑制式及特色—土楼的作用—土楼的建筑材料—田螺坑土楼群—二宜楼的建筑制式—民族文化精神解读		
计划制订的方式	同学间分组讨论		

序号	工作步骤	注意事项
1	查看资料文件	
2	查询资料	
3	设计课程内容思维导图	

计划评价	班级		第___组	组长签字	
	教师签字		日期		
	评语：				

决策单

学习领域	福建土楼		
学习情境	学习福建土楼的相关知识点		
学习任务	熟练掌握福建土楼群的名称、承启楼的建筑制式、田螺坑土楼群的组成部分及土楼的作用	学时	10分钟
典型工作过程描述	教学情境导入—福建土楼群的分布及名称—承启楼的建筑制式及特色—土楼的作用—土楼的建筑材料—田螺坑土楼群—二宜楼的建筑制式—民族文化精神解读		

学生为主体的课堂决策

序号	课程目标	课程内容	教学活动	教学资料	综合评价
1					
2					
3					
4					
5					
6					
7					
8					
9					
10					

决策评价	班级		第___组	组长签字	
	教师签字		日期		
	评语：				

实施单

学习领域	福建土楼		
学习情境	学习福建土楼的相关知识点		
学习任务	熟练掌握福建土楼群的名称、承启楼的建筑制式、田螺坑土楼群的组成部分及土楼的作用	学时	30分钟
典型工作过程描述	教学情境导入—福建土楼群的分布及名称—承启楼的建筑制式及特色—土楼的作用—土楼的建筑材料—田螺坑土楼群—二宜楼的建筑制式—民族文化精神解读		

序号	实施步骤	注意事项
1	教学情境导入	播放先导片
2	承启楼的建筑制式与建筑特色	教学重点与难点
3	土楼的建筑材料及作用	教学重点与难点
4	田螺坑土楼群组成部分名称	教学重点与难点
5	结合课程思政	引入中华民族传统文化精神

实施说明：
1. 播放先导片；
2. 借助微课视频与PPT进行深入了解；
3. 通过了解福建土楼引入中华民族传统文化精神

实施评价	班级		第___组	组长签字	
	教师签字		日期		
	评语：				

检查单

学习领域	福建土楼		
学习情境	学习福建土楼的相关知识点		
学习任务	熟练掌握福建土楼群的名称、承启楼的建筑制式、田螺坑土楼群的组成部分及土楼的作用	学时	15分钟
典型工作过程描述	教学情境导入—福建土楼群的分布及名称—承启楼的建筑制式及特色—土楼的作用—土楼的建筑材料—田螺坑土楼群—二宜楼的建筑制式—民族文化精神解读		

序号	检查项目	检查标准	学生自查	教师检查
1	资讯环节	获取相关信息情况		
2	计划环节	设计课程思维导图		
3	实施环节	绘制课程思维导图		
4	检查环节	各个环节逐一检查		

检查评价	班级		第___组	组长签字	
	教师签字		日期		
	评语：				

评价单

学习领域	福建土楼		
学习情境	学习福建土楼的相关知识点		
学习任务	熟练掌握福建土楼群的名称、承启楼的建筑制式、田螺坑土楼群的组成部分及土楼的作用	学时	15分钟
典型工作过程描述	教学情境导入—福建土楼群的分布及名称—承启楼的建筑制式及特色—土楼的作用—土楼的建筑材料—田螺坑土楼群—二宜楼的建筑制式—民族文化精神解读		

评价项目	评价子项目	学生自评	组内评价	教师评价
资讯环节	1. 听取教师讲解； 2. 互联网查询情况； 3. 同学交流情况			
计划环节	1. 查询资料情况； 2. 设计思维导图			
实施环节	1. 学习态度； 2. 课程内容熟练程度； 3. 课堂表现； 4. 作业完成情况及质量			
最终结果	综合情况			

	班级		第___组	组长签字	
评价	教师签字		日期		
	评语：				

16.2 理论指导

福建土楼

16.2.1 福建土楼概览

福建土楼在 2008 年被列入世界遗产名录。福建土楼，指福建省永定、南靖、华安三县的"六群四楼"共 46 座土楼。其中直径为 66 m 的集庆楼已有 600 岁的"高龄"，直径为 31 m 的善庆楼仅有 30 年历史。

福建土楼主要分为永定土楼、南靖土楼、华安土楼，其中永定土楼中最为著名的是承启楼，南靖土楼中最为著名的是田螺坑土楼，华安土楼中最为著名的是二宜楼。

16.2.2 承启楼的建筑制式及建筑特色

承启楼有"圆楼之王"的美誉，高 4 层，楼有 4 圈，上下共有 400 间，圆中圆，圈套圈，历经沧桑 300 年。

承启楼的直径为 73 m，走廊周长为 229.34 m，有楼梯 4 部、房间 400 个。外环楼高 4 层，每层用抬梁式木构架镶嵌泥砖分隔成 72 开间；二层外墙不开窗，只在内墙开一小窗，从天井采光；一层是灶房，也就是厨房，二层是禾仓，禾仓用于放置劳动工具与粮食作物；三、四层是卧室；各层都有一条内向挑出的环形通道，共有 4 道楼梯，对称分布于楼内 4 个方向。第二环楼有 2 层，每层有 40 个房间。第三环楼为单层，有 32 个房间，中心是祖堂。三环楼就像 3 员大将紧紧守护着祖堂。这就是土楼大王的威仪，庄重而又壮观（图 16–1、图 16–2）。

图 16–1　承启楼的外观

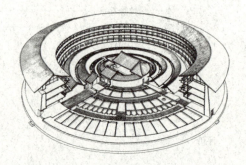
图 16–2　承启楼的内部结构

16.2.3 土楼的建筑材料及作用

建造土楼是为了抵御洪水及野兽的侵袭，以及抵御外来侵略者。

土楼特殊的建筑材料与土楼的作用息息相关，主要建筑材料包括黏土、杉木、石料、沙、石灰、竹片、瓦等辅助材料。通过夯土将泥土压实，这样可以使泥土结实、密度大且缝隙少，最终形成耐腐蚀、防火、防震的土楼。

16.2.4 田螺坑土楼

田螺坑土楼被称为"四菜一汤"。田螺坑土楼主要由1座方楼、3座圆楼及1座椭圆楼组成，分别是步云楼、文昌楼、振昌楼、瑞云楼、和昌楼，方楼步云楼居中，其余4座土楼环绕周围，这些土楼按"金木水火土"五行相生次序建造（图16-3）。

图16-3 田螺坑土楼

1. 步云楼

步云楼，取平步青云之意，寓意子孙后代读书中举，仕途步步高升，青云直上（图16-4、图16-5）。

图16-4 步云楼

图16-5 步云楼内部

2. 文昌楼

文昌楼于1966年建造，也是最后建造的一座土楼。它是目前仅存的一座椭圆形土楼，有3层高，每层有32个房间（图16-6）。

3. 振昌楼

振昌楼建于1930年，其与众不同之处是内堂与门不在同一直线上，反映了"富不露白"的文化理念。因为振昌楼侧后方有祖坟，于是把内堂设于祖坟前面，所以内堂方向是与大门方向错开的（图16-7、图16-8）。

图16-6 文昌楼

图16-7 振昌楼

图16-8 振昌楼内部

4. 瑞云楼

瑞云楼建于1936年,它坐落在5座土楼的内隅,有藏风聚气之功,体现了含蓄、吉顺的朴素观念(图16-9)。

5. 和昌楼

和昌楼有3层高,每层有22个房间,并设有2部楼梯(图16-10)。

图16-9 瑞云楼

图16-10 和昌楼

16.2.5 二宜楼

二宜楼前后建造了30年,占地面积为9 300 m²,直径为73.4 m,是福建省内同类建筑中单体最大的双环圆形土楼。内环平房为透天房屋,设有厨房、餐厅。外环楼每单元各4间。底层为卧室、客厅,第二、三层为卧室、仓库,第四层为各单元祖堂。

二宜楼外墙构造与永定土楼的承启楼外墙构造是相同的,外墙厚度是非常大的(图16-11)。

二宜楼的特色在于其大门防火系统,如果大门遭到火势攻击就可以通过竹筒注水来熄灭火源,这里充分体现了客家人的聪颖和智慧(图16-12)。

二宜楼的中心是作为公共场所的内院,在场地上可晾晒衣服和农作物,院中有两口古井,分别名为阴泉和阳泉,组成太极的阵形(图16-13)。

图16-11 二宜楼内部构造

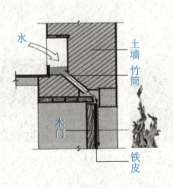

图16-12 二宜楼大门防火系统

图16-13 阴泉和阳泉

16.3　中华民族传统文化精神解读

　　土楼建筑理念上的天人合一、建筑形态上的宜人宜居、文化意蕴上的和谐灵动，彰显了中国传统民居建筑的独特风采与客家人的智慧，其生态型、环保型、开放型的建筑思路对现代建筑营造具有较大的启示意义。

16.4　项目创新

　　回顾本节课的内容，比较承启楼与二宜楼的异同，如可以从功能、构造、建筑材料、布局特点等方面进行比较，并形成PPT进行汇报。

项目 17　园林中的建筑元素

17.1　项目表单

学习任务单

学习领域	园林中的建筑元素					
学习情境	学习园林中的建筑元素的相关知识点					
学习任务	熟练掌握园林中建筑元素的特点、中国四大名亭及江南三大名楼				学时	2学时（90分钟）
教学目标	使学生充分了解构成园林的建筑元素的名称、特点及中国四大名亭与江南三大名楼所包含的建筑名称					
技能目标	通过园林中的建筑元素进一步了解中国古典园林结构的精髓以及中国优良的传统文化					
知识目标	熟练掌握园林中建筑元素的特点、中国四大名亭及江南三大名楼					
学习目标	1. 园林中的建筑元素包括哪些？ 2. 中国四大名亭包括哪些建筑？ 3. 江南三大名楼包括哪些建筑					
任务描述	熟练掌握园林中建筑元素的特点、中国四大名亭及江南三大名楼					
学时安排	情境导入	计划	决策	实施	检查	评价
时长	10分钟	10分钟	10分钟	30分钟	15分钟	15分钟
对学生的要求	1. 课前做好项目预习； 2. 做好课中总结与评价； 3. 通过园林中的建筑元素的学习内容谈谈古典园林建筑语言中的人文主义精神及中华民族传统文化					
参考资料	1. 微课视频； 2. PPT； 3. 课程相关视频资料					

项目 17　园林中的建筑元素　157

项目前学习任务单

学习领域	园林中的建筑元素		用时	
达成目标	1. 观看微课视频，可以简单整理出课程内容思维导图； 2. 根据课程内容，感受中国古建筑文化的博大精深及大国工匠精神			
学习方法	看：观看微课视频及相关课程资料； 思：认真思考任务单中的问题； 记：将心中的疑惑记下来			
课堂学习形式预告	课堂外		课堂内	
	1. 自主学习，搜集资料； 2. 整理收获，提出问题		1. 展示交流； 2. 协作探究； 3. 启发式教学； 4. 项目驱动； 5. 实习作业法	
本次学习任务	任务 1：园林中的建筑元素包括哪些？建筑特色是什么？ 任务 2：中国四大名亭包括哪些建筑？ 任务 3：江南三大名楼包括哪些建筑？			
疑点与问题	本次学习中还有哪些不明白的地方？请依次记录下来。			
总结评价	1. 你认为自己预习得怎么样？ 　A. 马马虎虎，只是为了完成任务 　B. 态度认真，收获很多 2. 你对自己本次自主学习的评价：☆☆☆☆☆			

资讯单

学习领域	园林中的建筑元素		
学习情境	学习园林中的建筑元素的相关知识点		
学习任务	熟练掌握园林中建筑元素的特点、中国四大名亭及江南三大名楼	学时	10分钟
资讯方式	1. 教师讲解； 2. 互联网查询； 3. 同学交流		
资讯问题	1. 园林中的建筑元素包括哪些？建筑特色是什么？ 2. 中国四大名亭包括哪些建筑？ 3. 江南三大名楼包括哪些建筑？ 4. 通过古建筑园林中的建筑元素的学习内容谈谈中国古典园林建筑语言中的人文主义精神及中华民族传统文化		
资讯引导	1. 微课视频； 2. PPT； 3. 课程相关视频资料		

计划单

学习领域	园林中的建筑元素		
学习情境	学习园林中的建筑元素的相关知识点		
学习任务	熟练掌握园林中建筑元素的特点、中国四大名亭及江南三大名楼	学时	10分钟
典型工作过程描述	教学情境导入—园林中建筑元素的名称与特点—中国四大名亭—江南三大名楼—民族文化精神解读		
计划制订的方式	同学间分组讨论		

序号	工作步骤	注意事项
1	查看资料文件	
2	查询资料	
3	设计课程内容思维导图	

计划评价	班级		第___组	组长签字	
	教师签字		日期		
	评语：				

项目 17　园林中的建筑元素

决策单

学习领域	园林中的建筑元素		
学习情境	学习园林中的建筑元素的相关知识点		
学习任务	熟练掌握园林中建筑元素的特点、中国四大名亭及江南三大名楼	学时	10分钟
典型工作过程描述	教学情境导入—园林中建筑元素的名称与特点—中国四大名亭—江南三大名楼—民族文化精神解读		

学生为主体的课堂决策

序号	课程目标	课程内容	教学活动	教学资料	综合评价
1					
2					
3					
4					
5					
6					
7					
8					
9					
10					

决策评价	班级		第___组	组长签字	
	教师签字		日期		
	评语：				

实施单

学习领域	园林中的建筑元素		
学习情境	学习园林中的建筑元素的相关知识点		
学习任务	熟练掌握园林中建筑元素的特点、中国四大名亭及江南三大名楼	学时	30分钟
典型工作过程描述	教学情境导入—园林中建筑元素的名称与特点—中国四大名亭—江南三大名楼—民族文化精神解读		

序号	实施步骤	注意事项
1	教学情境导入	播放先导片
2	园林中建筑元素的名称与特点	教学重点与难点
3	中国四大名亭	教学重点与难点
4	江南三大名楼	教学重点与难点
5	结合课程思政	引入中华民族传统文化精神

实施说明：
1. 播放先导片；
2. 借助微课视频与PPT进行深入了解；
3. 通过了解园林中的建筑元素引入中华民族传统文化精神

实施评价	班级		第___组	组长签字	
	教师签字		日期		
	评语：				

检查单

学习领域	园林中的建筑元素		
学习情境	学习园林中的建筑元素的相关知识点		
学习任务	熟练掌握园林中建筑元素的特点、中国四大名亭及江南三大名楼	学时	15分钟
典型工作过程描述	教学情境导入—园林中建筑元素的名称与特点—中国四大名亭—江南三大名楼—民族文化精神解读		

序号	检查项目	检查标准	学生自查	教师检查
1	资讯环节	获取相关信息情况		
2	计划环节	设计课程思维导图		
3	实施环节	绘制课程思维导图		
4	检查环节	各个环节逐一检查		

检查评价	班级		第___组	组长签字	
	教师签字		日期		
	评语：				

项目 17　园林中的建筑元素

评价单

学习领域	园林中的建筑元素			
学习情境	学习园林中的建筑元素的相关知识点			
学习任务	熟练掌握园林中建筑元素的特点、中国四大名亭及江南三大名楼		学时	15 分钟
典型工作过程描述	教学情境导入—园林中建筑元素的名称与特点—中国四大名亭—江南三大名楼—民族文化精神解读			
评价项目	评价子项目	学生自评	组内评价	教师评价
资讯环节	1. 听取教师讲解； 2. 互联网查询情况； 3. 同学交流情况			
计划环节	1. 查询资料情况； 2. 设计思维导图			
实施环节	1. 学习态度； 2. 课程内容熟练程度； 3. 课堂表现； 4. 作业完成情况及质量			
最终结果	综合情况			
评价	班级		第___组	组长签字
	教师签字		日期	
	评语：			

17.2 理论指导

17.2.1 中国古典园林中的建筑元素及其特点

古典园林中的建筑元素

中国古典园林中的建筑元素包括亭、台、楼、阁、轩、榭、廊、舫。这些字在英语中很难找到对应的词汇，都被笼统地称为 Pavilions 或者 Buildings，因为欧美人认为这些园林中的建筑元素是相近的，功能也基本相同。其实在古代建筑中，亭、台、楼、阁、轩、榭、廊、舫有着明确的定义和区分，其功能也根据所处的位置、规模的大小、形式的不同而有差异，观赏美感也不一样。

1. 亭

亭也称为凉亭，是一种中国传统建筑，源于周代，多建于路旁，供行人休息、乘凉或观景。亭一般为开敞性结构，没有围墙，建筑形态特点为有顶无墙，顶部可分为六角、八角、圆形等多种形状。亭因为造型轻巧、选材不拘、布设灵活而被广泛应用于园林建筑（图17-1）。

2. 台

台起源于商、周，盛行于春秋、战国时期，是最古老的园林建筑形式。古代的宫殿建筑多建于台之上。早期的台是一种高耸的夯土建筑，以作登眺之用，其功能主要是祭祀，兼有观景和娱乐功能。它后来演变成厅堂前的露天平台，即月台；还有戏台，主要供古时唱戏娱乐之用（图17-2）。

3. 楼

楼是古建筑中两层以上的房屋，也称重屋。楼在战国时就已出现，汉代城楼已高达3层（图17-3）。

图 17-1 亭

图 17-2 台

图 17-3 楼

4. 阁

人们经常说重屋为楼，四敞为阁。阁四面皆有窗，且设有门，四周还设有挑出的平座，供人环阁漫步、观景，也供藏书和供佛用。人们经常说"楼阁"，阁与楼似乎总是连着出现，但阁与楼相比，体量更小巧（图17-4）。

5. 轩

轩是指园林中小而明亮、适宜读书的建筑，也称为"斋"或"室"。轩的古意为有窗的长廊或小屋，多为高而敞的建筑，但体量不大。轩的形式类型较多，形状各异，如同宽的廊，是一种点缀性的建筑（图17-5）。

6. 榭

榭是建在高土台或水面上或临水的建筑，多指水榭。有榭的地方就会水，榭是供人们休息、观景的建筑。榭中多设栏杆和坐凳，这些栏杆与坐凳又称为美人靠、吴王靠，相传是吴越时吴王与美人西施游赏观景时休息所用（图17-6）。

图17-4 阁

图17-5 轩

图17-6 榭

7. 廊

廊是连接两个建筑物的通道。其上有顶棚，以柱支撑，用以遮阳、挡雨，便于人们在游走过程中观赏景物。从建筑艺术的角度来说，廊增加了空间层次（图17-7）。

8. 舫

舫是仿照船的造型，在园林的水面上建造的一种船形建筑物，舫就是船的意思。舫似船而不能划动，故又称为"不系舟"。舫大多三面临水，一面与陆地相连，供人们游玩设宴、观赏水景（图17-8）。

图17-7 廊

图17-8 舫

17.2.2 中国四大名亭

中国四大名亭分别是醉翁亭、陶然亭、爱晚亭、湖心亭。

1. 醉翁亭

醉翁亭是中国四大名亭之首，又称为"天下第一亭"，位于今安徽省滁州市西南的琅琅山风景名胜区（图17-9）。

2. 陶然亭

陶然亭位于北京，当时的工部郎中江藻奉命监理黑窑厂，他在慈悲庵西部构筑了一座小亭，并取白居易诗"更待菊黄家酿熟，与君一醉一陶然"句中的"陶然"二字为亭命名（图17-10）。

图17-9　醉翁亭

图17-10　陶然亭

3. 爱晚亭

爱晚亭位于湖南省长沙市，原名"红叶亭"，又名"爱枫亭"，因唐代诗人杜牧的诗作《山行》而改名为"爱晚亭"，取"停车坐爱枫林晚，霜叶红于二月花"之诗意（图17-11）。

爱晚亭的建筑制式是八柱重檐，顶部覆以绿色琉璃瓦，攒尖宝顶，内柱为红色木柱，外柱为花岗石方柱，内部是天花彩绘藻井。

4. 湖心亭

湖心亭位于浙江省杭州市西湖中央，在西湖十八景中被称为"湖心平眺"。在湖心亭极目四眺，湖光尽收眼底（图17-12）。

图17-11　爱晚亭

图17-12　湖心亭

17.2.3　江南三大名楼

中国的江南三大名楼包括湖北的黄鹤楼、湖南的岳阳楼及江西的滕王阁。

1. 黄鹤楼

黄鹤楼位于湖北省武汉市，现在已成为武汉市的地标性建筑，为中国江南三大名楼之首。黄鹤楼主楼为四边套八边形体，通高51.4 m，底层边宽30 m，顶层边宽18 m，有飞檐5层，攒尖楼顶，顶覆金色琉璃瓦，由72根圆柱支撑，楼上有60个翘角向外伸展。

黄鹤楼外观有 5 层，内部实际有 9 层，隐含"九五至尊"之意，八方飞檐的鹤翼造型体现了黄鹤楼的独特文化，使中国传统建筑特色与文化意蕴完美结合（图 17-13）。

2. 岳阳楼

岳阳楼主楼为长方形体，主楼高 19.42 m，进深 14.54 m，宽 17.42 m，为三层、四柱、飞檐、盔顶、纯木结构，楼中 4 根楠木金柱直贯楼顶，周围绕以廊、枋、椽、檩互相榫合，结为整体；顶覆琉璃黄瓦，构型庄重大方。岳阳楼内一楼悬挂《岳阳楼记》雕屏及诗文、对联、雕刻等；二楼正中悬挂紫檀木雕屏，上刻有清朝书法家张照书写的《岳阳楼记》；三楼悬有毛泽东手书的杜甫《登岳阳楼》诗词雕屏，檐柱上挂"长庚李白书"对联"水天一色，风月无边"，具有一定的观赏价值。岳阳楼作为三大名楼中唯一保持原貌的古建筑，其独特的盔顶结构体现了古代劳动人民的聪明智慧及能工巧匠的精巧设计技能（图 17-14）。

3. 滕王阁

滕王阁位于江西省南昌市，为南昌市的地标性建筑。滕王阁是唐太宗李世民之弟滕王李元婴任江南洪州都督时所修，世称"西江第一楼"。

滕王阁主体建筑高 57.5 m，建筑面积为 13 000 m²；其下部为象征古城墙的 12 m 高台座，分为两级；台座以上的主阁取"明三暗七"格式，三层带回廊仿宋式建筑，内部共有 7 层，分为 3 个明层、3 个暗层及阁楼（图 17-15）。

图 17-13　黄鹤楼

图 17-14　岳阳楼

图 17-15　滕王阁

17.3　中华民族传统文化精神解读

"天人合一"的理念是中国古典园林艺术的灵魂。其主要观点是在尊重自然的前提下改造自然，创造出宜人、和谐的园林生态。在工业文明威胁人类生存环境的当今世界，这种思想理念本身蕴含着某种解决的途径和哲学道理，很值得人们学习与借鉴，特别在现代规划和设计中可以继承这一伟大的思想。

17.4　项目创新

回顾本节课的内容，将所学到的"亭、台、楼、阁、轩、榭、廊、坊"的相关知识复习整理并以思维导图的形式呈现。

项目 18　园林组成要素

18.1　项目表单

学习任务单

学习领域	园林组成要素					
学习情境	学习园林组成要素的相关知识点					
学习任务	熟练掌握园林组成要素、叠山石材的选择标准及质地、叠山理水的功能			学时	2学时（90分钟）	
教学目标	使学生充分了解园林组成要素包括哪些内容，以及叠山石材的选择标准及质地，叠山理水的主要功能					
技能目标	通过园林组成要素进一步了解中国古典园林结构的精髓及中国优良的传统文化					
知识目标	熟练掌握园林组成要素、叠山石材的选择标准及质地、叠山理水的功能					
学习目标	1. 园林组成要素包括哪些？ 2. 叠山石材的选择标准及质地是什么？ 3. 叠山理水的功能分别是什么？					
任务描述	熟练掌握园林组成要素、叠山石材的选择标准及质地、叠山理水的功能					
学时安排	情境导入	计划	决策	实施	检查	评价
时长	10分钟	10分钟	10分钟	30分钟	15分钟	15分钟
对学生的要求	1. 课前做好项目预习； 2. 做好课中总结与评价； 3. 通过园林组成要素的学习内容谈谈古典园林建筑语言中的人文主义精神及中华民族传统文化					
参考资料	1. 微课视频； 2. PPT； 3. 课程相关视频资料					

项目 18　园林组成要素　167

项目前学习任务单

学习领域	园林组成要素		用时	
达成目标	1. 观看微课视频，可以简单整理出课程内容思维导图； 2. 根据课程内容，感受中国古建筑文化的博大精深及大国工匠精神			
学习方法	看：观看微课视频及相关课程资料； 思：认真思考任务单中的问题； 记：将心中的疑惑记下来			
课堂学习形式预告	课堂外		课堂内	
	1. 自主学习，搜集资料； 2. 整理收获，提出问题		1. 展示交流； 2. 协作探究； 3. 启发式教学； 4. 项目驱动； 5. 实习作业法	
本次学习任务	任务 1：园林组成要素包括哪些？ 任务 2：叠山石材的选择标准及质地是什么？ 任务 3：叠山理水的功能分别是什么？			
疑点与问题	本次学习中还有哪些不明白的地方？请依次记录下来。 _____ _____ _____ _____ _____ _____ _____ _____ _____			
总结评价	1. 你认为自己预习得怎么样？ 　A. 马马虎虎，只是为了完成任务 　B. 态度认真，收获很多 2. 你对自己本次自主学习的评价：☆☆☆☆☆			

资讯单

学习领域	园林组成要素		
学习情境	学习园林组成要素的相关知识点		
学习任务	熟练掌握园林组成要素、叠山石材的选择标准及质地、叠山理水的功能	学时	10分钟
资讯方式	1. 教师讲解； 2. 互联网查询； 3. 同学交流		
资讯问题	1. 园林组成要素包括哪些？ 2. 叠山石材的选择标准及质地是什么？ 3. 叠山理水的功能分别是什么？ 4. 通过古建筑园林组成要素的学习内容谈谈中国古典园林建筑语言中的人文主义精神及中华民族传统文化		
资讯引导	1. 微课视频； 2. PPT； 3. 课程相关视频资料		

计划单

学习领域	园林组成要素		
学习情境	学习园林组成要素的相关知识点		
学习任务	熟练掌握园林组成要素、叠山石材的选择标准及质地、叠山理水的功能	学时	10分钟
典型工作过程描述	教学情境导入—园林组成要素概览—叠山石材的选择标准—叠山石材的质地—叠山的作用—叠山理水的类型—叠山理水的功能—植物—动物—建筑—楹联匾额—刻石—民族文化精神解读		
计划制订的方式	同学间分组讨论		

序号	工作步骤	注意事项
1	查看资料文件	
2	查询资料	
3	设计课程内容思维导图	

计划评价	班级		第___组		组长签字	
	教师签字		日期			
	评语：					

决策单

学习领域	园林组成要素		
学习情境	学习园林组成要素的相关知识点		
学习任务	熟练掌握园林组成要素、叠山石材的选择标准及质地、叠山理水的功能	学时	10分钟
典型工作过程描述	教学情境导入—园林组成要素概览—叠山石材的选择标准—叠山石材的质地—叠山的作用—叠山理水的类型—叠山理水的功能—植物—动物—建筑—楹联匾额—刻石—民族文化精神解读		

学生为主体的课堂决策

序号	课程目标	课程内容	教学活动	教学资料	综合评价
1					
2					
3					
4					
5					
6					
7					
8					
9					
10					

决策评价	班级		第___组	组长签字	
	教师签字		日期		
	评语：				

实施单

学习领域	园林组成要素		
学习情境	学习园林组成要素的相关知识点		
学习任务	熟练掌握园林组成要素、叠山石材的选择标准及质地、叠山理水的功能	学时	30分钟
典型工作过程描述	教学情境导入—园林组成要素概览—叠山石材的选择标准—叠山石材的质地—叠山的作用—叠山理水的类型—叠山理水的功能—植物—动物—建筑—楹联匾额—刻石—民族文化精神解读		

序号	实施步骤	注意事项
1	教学情境导入	播放先导片
2	园林组成要素	教学重点与难点
3	叠山石材的选择标准及质地	教学重点与难点
4	叠山理水的功能	教学重点与难点
5	结合课程思政	引入"中华民族传统文化精神"

实施说明：
1. 播放先导片；
2. 借助微课视频与PPT进行深入了解；
3. 通过了解园林组成要素引入中华民族传统文化精神

实施评价	班级		第___组	组长签字	
	教师签字		日期		
	评语：				

检查单

学习领域	园林组成要素		
学习情境	学习园林组成要素的相关知识点		
学习任务	熟练掌握园林组成要素、叠山石材的选择标准及质地、叠山理水的功能	学时	15分钟
典型工作过程描述	教学情境导入—园林组成要素概览—叠山石材的选择标准—叠山石材的质地—叠山的作用—叠山理水的类型—叠山理水的功能—植物—动物—建筑—楹联匾额—刻石—民族文化精神解读		

序号	检查项目	检查标准	学生自查	教师检查
1	资讯环节	获取相关信息情况		
2	计划环节	设计课程思维导图		
3	实施环节	绘制课程思维导图		
4	检查环节	各个环节逐一检查		

检查评价	班级		第___组	组长签字	
	教师签字		日期		
	评语：				

项目 18　园林组成要素

评价单

学习领域	园林组成要素		
学习情境	学习园林组成要素的相关知识点		
学习任务	熟练掌握园林组成要素、叠山石材的选择标准及质地、叠山理水的功能	学时	15 分钟
典型工作过程描述	教学情境导入—园林组成要素概览—叠山石材的选择标准—叠山石材的质地—叠山的作用—叠山理水的类型—叠山理水的功能—植物—动物—建筑—楹联匾额—刻石—民族文化精神解读		

评价项目	评价子项目	学生自评	组内评价	教师评价
资讯环节	1. 听取教师讲解； 2. 互联网查询情况； 3. 同学交流情况			
计划环节	1. 查询资料情况； 2. 设计思维导图			
实施环节	1. 学习态度； 2. 课程内容熟练程度； 3. 课堂表现； 4. 作业完成情况及质量			
最终结果	综合情况			

	班级		第___组	组长签字	
	教师签字		日期		
评价	评语：				

18.2 理论指导

园林主要包括山、水、植物、动物、建筑、楹联、匾额、石刻八大构景要素。将它们按照中国传统艺术规律进行设计与组合,以反映园林艺术的自然精神境界为最终和最高目的,从而达到"虽由人作,宛自天开"的审美旨趣。

1. 叠山

叠山是取真山的山姿山容、气势风韵,经过艺术概括、提炼,再现在园林里,以小山之形传大山之神。这类以某种真山的意境创作而成的山体,会给人一种亲切感,有着丰富的想象和品位的余地(图 18-1)。

传统叠山的石材选石标准为透、瘦、皱、漏、丑。所谓透,即玲珑多孔,外形轮廓多姿;瘦者,瘦中寓秀;皱者,外形起伏不平,明暗多变化,富有美好的节奏感;漏则是石峰上下左右,窍窍相通,有路可循;清丑的石峰颇具气势,能创造出苍劲古朴的意境,引人遐想。

石材的质地可以分为以下几种。

湖石,原产于太湖一带,是园林中运用最为普遍,也是历史上开发较早的一类山石。其石面多孔,质坚而脆、纹理纵横。其子类有太湖石、房山石、英德石、灵璧石、黄石、石笋、宣石等。

黄石,是一种呈黄色的细砂岩,产地较多。其棱角分明、雄浑沉实。

石笋,其外形如修长的竹笋,产地广,皆卧于土山之中,开采出来后直立于地面。石笋常作为园林中的独立小景(图 18-2)。

其中太湖石和黄石是最为常见的石材。

图 18-1 叠山

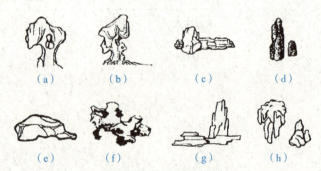

图 18-2 选石标准
(a)太湖石;(b)灵璧石;(c)黄石;(d)石笋;(e)黄蜡石;
(f)英德石;(g)宣石;(h)钟乳石

叠山在园林艺术中有三大功能。它首先可以组织划分、分隔空间;其次能因地制宜、协调环境;最后可以点缀风景。

2. 理水

理水泛指各类园林中水景处理。水是流动的、不定型的,与山的稳重、固定形成鲜明的对比。传统园林中的理水是对自然山水特征的概括、提炼和再现。各类水的形态的表现不在于绝对体量接近自然,而在于风景特征的艺术真实。

园林中的理水有泉瀑、渊潭、溪涧、河流、湖泊几种常见的类型。

泉为地下涌出的水,瀑是从断崖跌落的水,水源或为天然泉水,或为园外引水,或为人工水源。瀑布有线状、帘状、分流、叠落等形式。渊潭是指小而深的水体,光线幽暗,水位较低。溪涧是泉瀑之水从山间流出的一种动态水景,以砾石为底,溪水较浅,可数游鱼,又可涉水。河流在园林中常用狭长形的水池来表现,使景色富有变化。河流可长可短,可直可弯,有宽有窄,有收有放。河流多用土岸,配置适当的植物。湖泊是指成片汇聚的水面(图 18-3 ～图 18-7)。

图 18-3　泉瀑

图 18-4　渊潭

图 18-5　溪涧

图 18-6　河流

图 18-7　湖泊

园林中的理水不仅可以造景,也可以排洪蓄水、改善生态环境、调节局部环境的小气候及开展各类水上活动与游览。

3. 植物元素

植物是叠山理水不可缺少的因素,水景如果离开花木则无美感,有了花木才使山更青、水更秀。

首先要重视植物的多样性。自然界中的植物千姿百态、丰富多彩,本身具有很高的观赏价值。其次要布局合理、疏朗有致、单群结合。自然界中的植物并不都是群生的,也有孤生的。园林植物配置就有孤植、列植、片植、群植、混植多种方式,这样不仅可欣赏孤植的风姿,也可欣赏群植的华美。除此之外,应注意不同园林植物形态和色彩的合理搭配,应根据地形、地貌配植不同形态、色彩的植物,而且它们相互之间不能造成视角上的抵触。最后要注意园林植物自身的文化性应与周围环境融合,如与道路的结合,以及与建筑的结合(图 18-8、图 18-9)。

4. 动物元素

园林中的动物主要供人观赏,更好地展现出与自然环境的融合,产生一种灵动的氛围感。园林中的动物主要有鸟、鸭子、鱼等，其中鱼在园林动物元素中占据首要位置（图18-10）。

图18-8　园林植物（一）

图18-9　园林植物（二）

图18-10　园林中的鱼

5. 建筑元素

园林中的建筑元素包括亭、台、楼、阁、轩、榭、廊、舫。

6. 匾额与楹联

匾额与楹联主要悬挂在厅、堂、亭、榭之上。匾额横置门头或墙洞门上，在园林中多为景点的名称或对景色的称颂；楹联往往与匾额相配，挂在两边。

匾额与楹联不但能点缀堂榭，装饰门墙，在园林中往往表达了造园者或园主的思想感情，还可以丰富景观，唤起联想，增加诗情画意，起到画龙点睛的作用。匾额与楹联是中国传统园林的一个特色（图18-11、图18-12）。

图18-11　匾额与楹联（一）

图18-12　匾额与楹联（二）

7. 刻石

刻石以奇秀、浑圆、皱瘦、浑厚、挺拔、秀丽为特征，属于景观石、园林石的一种，是镌刻诗文、题写碑额的重要载体。刻石天然质朴的外形、质感和汉字、书法富有动感画意的线条有机结合，本身就寓意了人力和自然的统一（图18-13、图18-14）。

图18-13　刻石（一）

图18-14　刻石（二）

18.3　中华民族传统文化精神解读

中国园林组成要素中有山、水、植物、动物、建筑、楹联、匾额及石刻，它们铸就了中国建筑中的传奇，同时也是世界艺术的奇观、人类文明的重要遗产。它们蕴含了中国的民族文化精神，将自然理念与人文精神进行和谐统一。它们是中国古代劳动人民智慧和创造力的结晶，也是中国古代哲学思想、文化艺术的综合体现。

18.4　项目创新

回顾本节课的内容，谈一谈你最感兴趣的一个园林组成元素，并完成PPT汇报作业。

项目 19　皇家园林

19.1　项目表单

学习任务单

学习领域	皇家园林					
学习情境	学习皇家园林颐和园的相关知识点					
学习任务	熟练掌握皇家园林颐和园的作用、主要建筑元素及建筑元素的功能和作用			学时	2学时（90分钟）	
教学目标	使学生充分了解皇家园林颐和园的作用、园林中的主要建筑元素及建筑元素的功能和作用					
技能目标	通过皇家园林颐和园进一步了解中国古典园林结构的精髓及中国优良的传统文化					
知识目标	熟练掌握皇家园林颐和园的作用、主要建筑元素及建筑元素的功能和作用					
学习目标	1. 熟练掌握皇家园林颐和园的作用； 2. 熟练掌握颐和园中的主要建筑元素； 3. 熟练掌握颐和园中主要建筑元素的功能和作用					
任务描述	熟练掌握皇家园林颐和园的作用、主要建筑元素及建筑元素的功能和作用					
学时安排	情境导入	计划	决策	实施	检查	评价
时长	10分钟	10分钟	10分钟	30分钟	15分钟	15分钟
对学生的要求	1. 课前做好项目预习； 2. 做好课中总结与评价； 3. 通过皇家园林的学习内容谈谈皇家园林建筑语言中的人文主义精神及中华民族传统文化					
参考资料	1. 微课视频； 2. PPT； 3. 课程相关视频资料					

项目前学习任务单

学习领域	皇家园林		用时	
达成目标	1. 观看微课视频，可以简单整理出课程内容思维导图； 2. 根据课程内容，感受中国古建筑文化的博大精深及大国工匠精神			
学习方法	看：观看微课视频及相关课程资料； 思：认真思考任务单中的问题； 记：将心中的疑惑记下来			
课堂学习形式预告	课堂外		课堂内	
	1. 自主学习，搜集资料； 2. 整理收获，提出问题		1. 展示交流； 2. 协作探究； 3. 启发式教学； 4. 项目驱动； 5. 实习作业法	
本次学习任务	任务1：颐和园的作用是什么？ 任务2：颐和园中的主要建筑元素包括哪些？ 任务3：颐和园中主要建筑元素的功能和作用是什么？			
疑点与问题	本次学习中还有哪些不明白的地方？请依次记录下来。 _____ _____ _____ _____ _____ _____ _____ _____ _____ _____ _____ _____ _____			
总结评价	1. 你认为自己预习得怎么样？ 　A. 马马虎虎，只是为了完成任务 　B. 态度认真，收获很多 2. 你对自己本次自主学习的评价：☆☆☆☆☆			

资讯单

学习领域	皇家园林		
学习情境	学习皇家园林颐和园的相关知识点		
学习任务	熟练掌握皇家园林颐和园的作用、主要建筑元素及建筑元素的功能和作用	学时	10分钟
资讯方式	1. 教师讲解； 2. 互联网查询； 3. 同学交流		
资讯问题	1. 颐和园的作用是什么？ 2. 颐和园中的主要建筑元素包括哪些？ 3. 颐和园中主要建筑元素的功能和作用是什么？ 4. 通过古建筑皇家园林的学习内容谈谈中国古典园林建筑语言中的人文主义精神及中华民族传统文化		
资讯引导	1. 微课视频； 2. PPT； 3. 课程相关视频资料		

计划单

学习领域	皇家园林		
学习情境	学习皇家园林颐和园的相关知识点		
学习任务	熟练掌握皇家园林颐和园的作用、主要建筑元素及建筑元素的功能和作用	学时	10分钟
典型工作过程描述	教学情境导入—颐和园概览—颐和园的功能和作用—颐和园的建筑布局—颐和园中的主要建筑元素—颐和园中主要建筑元素的功能和作用—民族文化精神解读		
计划制订的方式	同学间分组讨论		

序号	工作步骤	注意事项
1	查看资料文件	
2	查询资料	
3	设计课程内容思维导图	

计划评价	班级		第___组	组长签字	
	教师签字		日期		
	评语：				

决策单

学习领域	皇家园林		
学习情境	学习皇家园林颐和园的相关知识点		
学习任务	熟练掌握皇家园林—颐和园的作用、主要建筑元素及建筑元素的功能和作用	学时	10分钟
典型工作过程描述	教学情境导入—颐和园概览—颐和园的功能和作用—颐和园的建筑布局—颐和园中的主要建筑元素—颐和园中主要建筑元素的功能和作用—民族文化精神解读		

学生为主体的课堂决策					
序号	课程目标	课程内容	教学活动	教学资料	综合评价
1					
2					
3					
4					
5					
6					
7					
8					
9					
10					

决策评价	班级		第___组	组长签字	
	教师签字		日期		
	评语：				

实施单

学习领域	皇家园林		
学习情境	学习皇家园林颐和园的相关知识点		
学习任务	熟练掌握皇家园林颐和园的作用、主要建筑元素及建筑元素的功能和作用	学时	30分钟
典型工作过程描述	教学情境导入—颐和园概览—颐和园的功能和作用—颐和园的建筑布局—颐和园中的主要建筑元素—颐和园中主要建筑元素的功能和作用—民族文化精神解读		
序号	实施步骤	注意事项	
1	教学情境导入	播放先导片	
2	颐和园概览	教学重点与难点	
3	颐和园的功能和作用	教学重点与难点	
4	颐和园中主要建筑元素的功能和作用	教学重点与难点	
5	结合课程思政	引入中华民族传统文化精神	
实施说明： 1. 播放先导片； 2. 借助微课视频与PPT进行深入了解； 3. 通过了解皇家园林颐和园引入中华民族传统文化精神			
实施评价	班级	第___组	组长签字
	教师签字	日期	
	评语：		

检查单

学习领域	皇家园林		
学习情境	学习皇家园林颐和园的相关知识点		
学习任务	熟练掌握皇家园林颐和园的作用、主要建筑元素及建筑元素的功能和作用	学时	15分钟
典型工作过程描述	教学情境导入—颐和园概览—颐和园的功能和作用—颐和园的建筑布局—颐和园中的主要建筑元素—颐和园中主要建筑元素的功能和作用—民族文化精神解读		
序号	检查项目	检查标准	学生自查	教师检查
1	资讯环节	获取相关信息情况		
2	计划环节	设计课程思维导图		
3	实施环节	绘制课程思维导图		
4	检查环节	各个环节逐一检查		
检查评价	班级		第___组	组长签字
	教师签字		日期	
	评语：			

评价单

学习领域	皇家园林		
学习情境	学习皇家园林颐和园的相关知识点		
学习任务	熟练掌握皇家园林颐和园的作用、主要建筑元素及建筑元素的功能和作用	学时	15分钟
典型工作过程描述	教学情境导入—颐和园概览—颐和园的功能和作用—颐和园的建筑布局—颐和园中的主要建筑元素—颐和园中主要建筑元素的功能和作用—民族文化精神解读		

评价项目	评价子项目	学生自评	组内评价	教师评价
资讯环节	1. 听取教师讲解； 2. 互联网查询情况； 3. 同学交流情况			
计划环节	1. 查询资料情况； 2. 设计思维导图			
实施环节	1. 学习态度； 2. 课程内容熟练程度； 3. 课堂表现； 4. 作业完成情况及质量			
最终结果	综合情况			

评价	班级		第___组	组长签字	
	教师签字		日期		
	评语：				

19.2 理论指导

19.2.1 皇家园林概览

皇家园林

颐和园是我国清朝时期的皇家园林，前身为清漪园，坐落在北京西郊，全园占地面积为 3.009 km²，水面约占四分之三。

19.2.2 颐和园的功能及主要建筑布局

颐和园是皇家生活环境的一个重要组成部分，是供皇帝游憩或处理政务的场所。颐和园主要由万寿山和昆明湖两部分组成。它包括各种形式的宫殿园林建筑 3 000 余座，大致可分为行政、生活、游览 3 个部分。

19.2.3 颐和园中的主要建筑元素及功能和作用

1. 苏州街

苏州街又称为"买卖街"，是仿江南水镇而建。清漪园时期岸上有各式店铺，如玉器古玩店、绸缎店、点心铺、茶楼、金银首饰楼等。店铺中的店员都是太监、宫女装扮。店铺在皇帝游幸时开始"营业"。乾隆皇帝第一次巡幸江南时，因留恋江南苏州热闹的街肆铺面及物产风俗，所以命随行画师按照此形式绘制图式，同时将市井文化融入皇家建筑，这里面的"市井"指的是平民百姓（图 19-1）。

2. 万寿山

万寿山为燕山余脉，高 58.59 m，海拔为 108.94 m。从山脚的"云辉玉宇"牌楼，经排云门、二宫门、排云殿、德辉殿、佛香阁，直至山顶的智慧海，形成了一条层层上升的中轴线（图 19-2）。

图 19-1 苏州街

图 19-2 万寿山

3. 四大部洲

颐和园四大部洲是根据古印度哲学观念进行建造的。它是颐和园万寿山后山中部的一组汉藏建筑群，占地面积为 2 万 m²。清朝时期的皇室都信奉佛教（特别是晚晴时期），因此修建此场所。

四大部洲指的是佛教中的四大部洲，是梵文释义，包括北俱卢洲、南瞻部洲、东胜神洲、西牛贺洲，依次为方形、三角形、半月形、圆形4种不同的形状，又对应佛家称为"四大"的地、火、风、水（图19-3）。

四大部洲中的主要建筑有经幢和喇嘛塔。经幢是佛教装饰，上面有许多小佛像，还刻有金刚经，高3 m；四大部洲的南、西南、东北、西北有代表佛经"四智"的红、白、黑、绿4座喇嘛塔（图19-4～图19-8）。

黑塔为"平等性智"，意为佛视世界万物平等无差别；白塔为"大圆境智"，意为佛能清楚地认识世界万象；红塔为"妙观察智"，表示佛能明察善恶，妙观万法；绿塔为"成所作智"，表示信佛能成就自利和利他事业。

图19-3 四大部洲

图19-4 经幢

图19-5 黑塔

图19-6 白塔

图19-7 红塔

图19-8 绿塔

4. 佛香阁

佛香阁是一座八面三层四重檐的建筑；阁高41 m，阁内有8根巨大铁梨木擎天柱，结构复杂，为古典建筑精品。皇室在此做法事（图19-9）。

5. 宝云阁

宝云阁俗称铜殿或铜亭。其外形仿照木结构建筑的样式，重檐歇山顶。宝云阁通高7.55 m，重约207 t。其构件如柱、梁、椽、瓦、脊吻兽，连匾额等都像木结构。其坐落在一个汉白玉雕砌的须弥座上（图19-10）。

6. 排云殿

排云殿原是乾隆为其庆贺母亲六十寿辰而建的大报恩延寿寺，慈禧太后重建的时候更名为排云殿。"排云"二字取自晋代郭璞《游仙诗》——"神仙排云出，但见金银台"，整首诗用夸张的方法描绘大海中神仙的居处（图19-11）。

图 19-9　佛香阁

图 19-10　宝云阁

7. 长廊

长廊位于万寿山南麓，全长 728 m，是中国园林中最长的游廊。长廊还是一条五光十色的画廊，廊间的每根枋梁上都绘有彩画，共 14 000 余幅，色彩鲜明，富丽堂皇。彩画是中国木结构建筑上的装饰艺术。它的特点是：主要画面被括在大半圆的括线内（称为"包袱"）；无固定结构，全凭画工发挥，由同一题材可创作出不相同的画面。长廊彩画题材广泛，山林、花鸟、景物、人物均可入画。其中最引人入胜的当数人物故事画。长廊人物故事画共 2 000 多幅，没有任何两幅是相同的，画师们将中华数千年的历史文化浓缩在长廊子上（图 19-12 ～ 图 19-15）。

图 19-11　排云殿

图 19-12　长廊

图 19-13　长廊中的苏式彩画（一）

图 19-14　长廊中的苏式彩画（二）

图 19-15　长廊中的苏式彩画（三）

8. 大戏楼

大戏楼位于颐和园的德和园内,是中国现存最大的古戏楼。它高 21 m,分上、中、下 3 层。下层天花板中心有天井与上层戏台串通,中层戏台设有绞车,可巧设机关布景,上天入地,变化无穷。修建大戏楼是因为慈禧太后嗜好京剧(图 19-16)。

9. 清晏舫

清晏舫是一条大大的石船,也称石舫,全长 36 m。清晏舫用大理石雕刻堆砌而成,是颐和园内著名的水上建筑珍品,也是颐和园中唯一带有西洋风格的建筑。船身上建有两层船楼,船底花砖铺地,窗户上装有彩色玻璃,顶部以砖雕装饰(图 19-17)。

图 19-16　大戏楼

图 19-17　清晏舫

10. 铜牛

铜牛是一座建筑雕塑,位于十七孔桥东桥头北侧,于乾隆二十年(1755 年)用铜铸造,称为"金牛"。它的作用是镇压水患,它是颐和园中独特的人文景观和艺术珍品(图 19-18)。

11. 十七孔桥

十七孔桥是古代桥梁建筑的杰作。十七孔桥由 17 个桥洞组成,长 150 m,每个桥栏的望柱上都雕有神态各异的狮子,大小共 544 个。整个十七孔桥具有极高的美学价值、学术价值和使用价值(图 19-19)。

图 19-18　铜牛

图 19-19　十七孔桥

19.3　中华民族传统文化精神解读

颐和园集传统造园艺术之大成,借景周围的山水环境,既有皇家园林恢宏富丽的气势,又充满自然之趣，高度体现了中国园林"虽由人作，宛自天开"的造园准则。1998年12月，颐和园被联合国教科文组织列入世界遗产名录，并给予如下评价：北京的颐和园是对中国风景园林造园艺术一种杰出的展现，将人造景观与大自然和谐地融为一体；颐和园是中国的造园思想和实践的集中体现，这种思想和实践对整个东方园林艺术文化形式的发展起了关键性的作用；以颐和园为代表的中国皇家园林是中国名列世界四大文明古国之一的有力象征。

19.4　项目创新

回顾本节课的内容，写出对皇家园林颐和园的感想。

项目 20　私家园林

20.1　项目表单

学习任务单

学习领域	私家园林					
学习情境	学习私家园林的相关知识点					
学习任务	熟练掌握私家园林的建筑特点、中国四大园林的名称、园林中的构景手法及园林中的植物			学时	2学时（90分钟）	
教学目标	使学生充分了解私家园林的建筑特点、中国四大园林的名称及园林中的构景手法					
技能目标	通过私家园林拙政园进一步了解中国古典私家园林结构的精髓及中国优良的传统文化					
知识目标	熟练掌握私家园林的建筑特点、中国四大园林的名称及园林中的构景手法					
学习目标	1. 熟练掌握私家园林的建筑特点； 2. 熟练掌握中国四大园林的名称； 3. 熟练掌握拙政园（园林）的构景手法					
任务描述	熟练掌握私家园林的建筑特点、中国四大园林的名称及园林中的构景手法					
学时安排	情境导入	计划	决策	实施	检查	评价
时长	10分钟	10分钟	10分钟	30分钟	15分钟	15分钟
对学生的要求	1. 课前做好项目预习； 2. 做好课中总结与评价； 3. 通过私家园林拙政园的学习内容谈谈私家园林建筑语言中的人文主义精神及中华民族传统文化					
参考资料	1. 微课视频； 2. PPT； 3. 课程相关视频资料					

项目前学习任务单

学习领域	私家园林		用时	
达成目标	1. 观看微课视频，可以简单整理出课程内容思维导图； 2. 根据课程内容，感受中国古建筑文化的博大精深及大国工匠精神			
学习方法	看：观看微课视频及相关课程资料； 思：认真思考任务单中的问题； 记：将心中的疑惑记下来			
课堂学习形式预告	课堂外		课堂内	
	1. 自主学习，搜集资料； 2. 整理收获，提出问题		1. 展示交流； 2. 协作探究； 3. 启发式教学； 4. 项目驱动； 5. 实习作业法	
本次学习任务	任务1：私家园林的建筑特点是什么？ 任务2：中国四大园林包括哪些？ 任务3：拙政园（园林）的构景手法有哪些？			
疑点与问题	本次学习中还有哪些不明白的地方？请依次记录下来。			

总结评价	1. 你认为自己预习得怎么样？
	   A. 马马虎虎，只是为了完成任务
	   B. 态度认真，收获很多
	2. 你对自己本次自主学习的评价：☆☆☆☆☆			

资讯单

学习领域	私家园林		
学习情境	学习私家园林的相关知识点		
学习任务	熟练掌握私家园林的建筑特点、中国四大园林的名称、园林中的构景手法及园林中的植物	学时	10分钟
资讯方式	1. 教师讲解； 2. 互联网查询； 3. 同学交流		
资讯问题	1. 私家园林的建筑特点是什么？ 2. 中国四大园林包括哪些？ 3. 拙政园（园林）的构景手法有哪些？ 4. 通过古建筑私家园林的学习内容谈谈中国古典园林建筑语言中的人文主义精神及中华民族传统文化		
资讯引导	1. 微课视频； 2. PPT； 3. 课程相关视频资料		

计划单

学习领域	私家园林		
学习情境	学习私家园林的相关知识点		
学习任务	熟练掌握私家园林的建筑特点、中国四大园林的名称、园林中的构景手法及园林中的植物	学时	10分钟
典型工作过程描述	教学情境导入—私家园林的建筑特点—中国四大名园—园林中的构景手法—园林中的植物—民族文化精神解读		
计划制订的方式	同学间分组讨论		

序号	工作步骤	注意事项
1	查看资料文件	
2	查询资料	
3	设计课程内容思维导图	

计划评价	班级		第___组	组长签字	
	教师签字		日期		
	评语：				

决策单

学习领域	私家园林		
学习情境	学习私家园林的相关知识点		
学习任务	熟练掌握私家园林的建筑特点、中国四大园林的名称、园林中的构景手法及园林中的植物	学时	10分钟
典型工作过程描述	教学情境导入—私家园林的建筑特点—中国四大园林—园林中的构景手法—园林中的植物—民族文化精神解读		

学生为主体的课堂决策

序号	课程目标	课程内容	教学活动	教学资料	综合评价
1					
2					
3					
4					
5					
6					
7					
8					
9					
10					

决策评价	班级		第___组	组长签字	
	教师签字		日期		
	评语：				

项目 20　私家园林

实施单

学习领域	私家园林		
学习情境	学习私家园林的相关知识点		
学习任务	熟练掌握私家园林的建筑特点、中国四大园林的名称、园林中的构景手法及园林中的植物	学时	30分钟
典型工作过程描述	教学情境导入—私家园林的建筑特点—中国四大园林—园林中的构景手法—园林中的植物—民族文化精神解读		
序号	实施步骤	注意事项	
1	教学情境导入	播放先导片	
2	私家园林的建筑特点	教学重点与难点	
3	中国四大名园	教学重点与难点	
4	园林中的构景手法及园林中的植物	教学重点与难点	
5	结合课程思政	引入中华民族传统文化精神	

实施说明：
1. 播放先导片；
2. 借助微课视频与 PPT 深入了解；
3. 通过了解私家园林引入中华民族传统文化精神

实施评价	班级		第___组	组长签字	
	教师签字		日期		
	评语：				

检查单

学习领域	私家园林		
学习情境	学习私家园林的相关知识点		
学习任务	熟练掌握私家园林的建筑特点、中国四大园林的名称、园林中的构景手法及园林中的植物	学时	15分钟
典型工作过程描述	教学情境导入—私家园林的建筑特点—中国四大园林—园林中的构景手法—园林中的植物—民族文化精神解读		
序号	检查项目	检查标准	学生自查	教师检查
1	资讯环节	获取相关信息情况		
2	计划环节	设计课程思维导图		
3	实施环节	绘制课程思维导图		
4	检查环节	各个环节逐一检查		

检查评价	班级		第___组	组长签字	
	教师签字		日期		
	评语：				

评价单

学习领域	私家园林			
学习情境	学习私家园林的相关知识点			
学习任务	熟练掌握私家园林的建筑特点、中国四大园林的名称、园林中的构景手法及园林中的植物		学时	15分钟
典型工作过程描述	教学情境导入—私家园林的建筑特点—中国四大园林—园林中的构景手法—园林中的植物—民族文化精神解读			
评价项目	评价子项目	学生自评	组内评价	教师评价
资讯环节	1. 听取教师讲解； 2. 互联网查询情况； 3. 同学交流情况			
计划环节	1. 查询资料情况； 2. 设计思维导图			
实施环节	1. 学习态度； 2. 课程内容熟练程度； 3. 课堂表现； 4. 作业完成情况及质量			
最终结果	综合情况			
评价	班级		第___组	组长签字	
	教师签字		日期		
	评语：				

20.2 理论指导

20.2.1 私家园林的建筑特点

私家园林

私家园林的建筑特点如下。第一是规模较小。私家园林的占地面积一般只有几亩至十几亩，小者仅一亩或半亩。造园家的主要构思是"小中见大"，即在有限的范围内运用含蓄、扬抑、曲折、暗示等手法来启动人的主观再创造，曲折有致，营造一种似乎深邃不尽的景境，增强人们对于实际空间的感受。第二是水面建设。私家园林大多以水面为中心，四周散布建筑，构成一个个景点，几个景点围合而成景区。第三是修身养性，私家园林的主要功能是修身养性，闲适自娱。第四是清高风雅。私家园林的园主多是文人学士出身，而非建造师或专业人士，文人学士能诗会画，善于品评，园林风格以清高风雅、淡素脱俗为最高追求，充溢着浓郁的书卷气。第五是造园手法在布局上采用灵活多变的手法，善于仿造自然山水的形象，十分讲究园林的细部处理。

20.2.2 中国四大名园

苏州拙政园、北京颐和园、承德避暑山庄、苏州留园一起被誉为"中国四大名园"。

20.2.3 园林中的构景手法

1. 借景

借景是古典园林建筑中常用的构景手法之一，即在视力所及的范围内，将好的景色组织到园林视线中。通过借景，园林中的建筑、水面、山石、植物、动物交相辉映（图20-1）。

2. 添景

添景是指在空间较空旷、景观较单调、景深层次缺乏之地，添置某景以改变此状况的构景手法，主要起过渡作用。其相当于在远景与近景之间添加一个中景（图20-2）。

3. 框景

框景就是用有限的空间去采收无限空间局部画面的构景手法。如古代园林中，常用门、窗、洞、亭、柱、檐、栏等构成框界构景，从而达到突出局部景观或步移景异的效果（图20-3）。

图20-1 借景

图20-2 添景

图20-3 框景

4. 漏景

漏景是从框景发展而来。框景景色全观，漏景若隐若现，含蓄雅致。漏景可以用漏窗、漏墙、漏屏风、疏林等手法（图20-4）。

5. 对景

对景是指使两个景致相隔一定的空间，从而可以使观景者从此地观赏彼地景致的一种方法，如两个亭子遥相辉映（图20-5）。

6. 抑景

抑景的种类有山抑、树抑、曲抑。中国传统艺术历来讲究含蓄，忌一览无余、一望到边。园林造景也绝不会让人一走进门口就看到最好的景色，最好的景色往往藏在后面，这叫作"先藏后露""欲扬先抑"。如园林入口处常迎门挡以假山，这种处理叫作山抑，入口处有树遮挡，则叫作树抑（图20-6）。

图20-4 漏景

图20-5 对景

图20-6 抑景

20.2.4 园林中的植物

1. 竹子

竹子被喻为有气节的君子，象征坚贞、高风亮节、虚心向上（图20-7）。

2. 梅花

梅花象征高洁、坚强、谦虚的品格，给人以发奋图强的励志精神（图20-8）。

3. 银杏

银杏的寓意为长寿、坚韧、沉着，同时还代表永恒的爱（图20-9）。

图20-7 竹子

图20-8 梅花

图20-9 银杏

4. 松树

松树是生命力极强的常青树，象征意志坚强、坚贞不屈的品格，同时也是长寿的象征（图20-10）。

5. 菊花

菊花凌霜盛开，一身傲骨，象征高尚、坚强的情操（图 20-11）。

图 20-10　松树

图 20-11　菊花

6. 莲花

莲花"出淤泥而不染，濯清涟而不妖，中通外直"，把莲花喻为君子，象征圣洁（图 20-12）。

7. 石榴树

石榴树的寓意为多子多福、儿孙满堂（图 20-13）。

图 20-12　莲花

图 20-13　石榴树

20.3　中华民族传统文化精神解读

　　中国古典园林经过历代匠师的创造，以人工设计建造天然优美的景物，点缀环境，形成了独具风格的传统文化艺术特色。借景、添景、框景、漏景、对景、抑景的造园手法可谓匠心独运，以文人雅士为中心建造的园林艺术在中国古建筑史上留下了浓墨重彩的一笔，人文精神与人文情怀的融入也使园林艺术犹如一缕青墨，散发着独韵幽香。每一种植物的寓意与园林的意境融合，使园林与自然界浑然天成。

20.4　项目创新

　　回顾本节课的内容，谈一谈苏州园林所体现的文人造园风格。

项目 21　中南海

21.1　项目表单

学习任务单

学习领域	中南海		
学习情境	学习中南海的相关知识点		
学习任务	熟练掌握中南海中的建筑、瀛台中所发生的历史事件及中华人民共和国成立的过程	学时	2学时（90分钟）
教学目标	使学生充分了解中南海建筑的前世今生及中华人民共和国成立的过程		
技能目标	通过中南海进一步了解戊戌变法等历史事件及中华人民共和国成立的过程		
知识目标	熟练掌握中南海中的建筑、瀛台中所发生的历史事件及中华人民共和国成立的过程		
学习目标	1. 熟练掌握中南海建筑概览； 2. 熟练掌握戊戌变法等历史事件； 3. 熟练掌握中华人民共和国成立的过程		
任务描述	熟练掌握中南海中的建筑、瀛台中所发生的历史事件及中华人民共和国成立的过程		

学时安排	情境导入	计划	决策	实施	检查	评价
时长	10分钟	10分钟	10分钟	30分钟	15分钟	15分钟

对学生的要求	1. 课前做好项目预习； 2. 做好课中总结与评价； 3. 通过中南海的学习内容谈谈戊戌变法等历史事件及对中华人民共和国成立的感想。
参考资料	1. 微课视频； 2. PPT； 3. 课程相关视频资料

项目前学习任务单

学习领域	中南海		用时	
达成目标	1. 观看微课视频，可以简单整理出课程内容思维导图； 2. 根据课程内容，感受中华人民共和国成立的伟大时刻			
学习方法	看：观看微课视频及相关课程资料； 思：认真思考任务单中的问题； 记：将心中的疑惑记下来			
课堂学习形式预告	课堂外		课堂内	
	1. 自主学习，搜集资料； 2. 整理收获，提出问题		1. 展示交流； 2. 协作探究； 3. 启发式教学； 4. 项目驱动； 5. 实习作业法	
本次学习任务	任务 1：中南海中的主要建筑； 任务 2：戊戌变法等历史事件始末； 任务 3：中华人民共和国成立的过程			
疑点与问题	本次学习中还有哪些不明白的地方？请依次记录下来。 			
总结评价	1. 你认为自己预习得怎么样？ 　A. 马马虎虎，只是为了完成任务 　B. 态度认真，收获很多 2. 你对自己本次自主学习的评价：☆☆☆☆☆			

资讯单

学习领域	中南海		
学习情境	学习中南海的相关知识点		
学习任务	熟练掌握中南海中的建筑、瀛台中所发生的历史事件及中华人民共和国成立的过程	学时	10分钟
资讯方式	1. 教师讲解； 2. 互联网查询； 3. 同学交流		
资讯问题	1. 中南海中的主要建筑有哪些？ 2. 简述戊戌变法历史事件。 3. 简述中华人民共和国成立的过程。 4. 通过古建筑中南海的学习内容谈谈戊戌变法等历史事件及对中华人民共和国成立的感想		
资讯引导	1. 微课视频； 2. PPT； 3. 课程相关视频资料		

计划单

学习领域	中南海		
学习情境	学习中南海的相关知识点		
学习任务	熟练掌握中南海中的建筑、瀛台中所发生的历史事件及中华人民共和国成立的过程	学时	10分钟
典型工作过程描述	教学情境导入—中南海中的主要建筑—瀛台中所发生的历史事件—中华人民共和国成立的过程		
计划制订的方式	同学间分组讨论		

序号	工作步骤	注意事项
1	查看资料文件	
2	查询资料	
3	设计课程内容思维导图	

计划评价	班级		第___组	组长签字	
	教师签字		日期		
	评语：				

项目 21　中南海

决策单

学习领域	中南海		
学习情境	学习中南海的相关知识点		
学习任务	熟练掌握中南海中的建筑、瀛台中所发生的历史事件及中华人民共和国成立的过程	学时	10分钟
典型工作过程描述	教学情境导入—中南海中的主要建筑—瀛台中所发生的历史事件—中华人民共和国成立的过程		
学生为主体的课堂决策			

序号	课程目标	课程内容	教学活动	教学资料	综合评价
1					
2					
3					
4					
5					
6					
7					
8					
9					
10					

决策评价	班级		第___组	组长签字	
	教师签字		日期		
	评语：				

实施单

学习领域	中南海		
学习情境	学习中南海的相关知识点		
学习任务	熟练掌握中南海中的建筑、瀛台中所发生的历史事件及中华人民共和国成立的过程	学时	30分钟
典型工作过程描述	教学情境导入—中南海中的主要建筑—瀛台中所发生的历史事件—中华人民共和国成立的过程		

序号	实施步骤	注意事项
1	教学情境导入	播放先导片
2	中南海中的主要建筑	教学重点与难点
3	戊戌变法等历史事件	教学重点与难点
4	中华人民共和国成立的过程	教学重点与难点
5	结合课程思政	中华人民共和国成立

实施说明：
1. 播放先导片；
2. 借助微课视频与PPT进行深入了解；
3. 通过了解中南海引入戊戌变法等历史事件及中华人民共和国成立的过程

实施评价	班级		第___组	组长签字	
	教师签字		日期		
	评语：				

检查单

学习领域	中南海		
学习情境	学习中南海的相关知识点		
学习任务	熟练掌握中南海中的建筑、瀛台中所发生的历史事件及中华人民共和国成立的过程	学时	15分钟
典型工作过程描述	教学情境导入—中南海中的主要建筑—瀛台中所发生的历史事件—中华人民共和国成立的过程		

序号	检查项目	检查标准	学生自查	教师检查
1	资讯环节	获取相关信息情况		
2	计划环节	设计课程思维导图		
3	实施环节	绘制课程思维导图		
4	检查环节	各个环节逐一检查		

检查评价	班级		第___组	组长签字	
	教师签字		日期		
	评语：				

评价单

学习领域	中南海		
学习情境	学习中南海的相关知识点		
学习任务	熟练掌握中南海中的建筑、瀛台中所发生的历史事件及中华人民共和国成立的过程	学时	15分钟
典型工作过程描述	教学情境导入—中南海中的主要建筑—瀛台中所发生的历史事件—中华人民共和国成立的过程		
评价项目	评价子项目	学生自评	组内评价	教师评价
---	---	---	---	---
资讯环节	1. 听取教师讲解； 2. 互联网查询情况； 3. 同学交流情况			
计划环节	1. 查询资料情况； 2. 设计思维导图			
实施环节	1. 学习态度； 2. 课程内容熟练程度； 3. 课堂表现； 4. 作业完成情况及质量			
最终结果	综合情况			

评价	班级		第___组	组长签字	
	教师签字		日期		
	评语：				

21.2 理论指导

中南海

21.2.1 中南海建筑概览

中南海，古称西苑、太液池，位于中国北京市西城区西长安街北侧，地处故宫和景山西侧，是中海和南海的统称，与北海合称"三海"。

中南海总占地面积约 100 万 m^2，其中水域面积约 46 万 m^2，主要由紫光阁、勤政殿、仪鸾殿、蕉园、水云榭、瀛台、丰泽园、静谷等建筑及园林组成，既有庄严肃穆的宫殿式建筑，又有灵活多变的园林式建筑，是中国现存较完整的皇家宫苑之一（图 21-1）。

中南海始建于辽代，在明清时期成为皇家御苑，现为中华人民共和国国务院、中共中央书记处和中共中央办公厅等重要机关办公地，是北京市的地标性建筑、中国国家象征之一。

瀛台是位于中南海南海中的仙岛皇宫，始建于明朝，清朝顺治、康熙年间曾两次修建，是帝王、后妃的听政、避暑和居住之地。其因四面临水，衬以亭台楼阁，像一座海中仙岛，故名瀛台（图 21-2）。

瀛台与著名的历史事件——戊戌变法有着千丝万缕的联系，戊戌变法失败后，慈禧太后将光绪帝囚禁于此。戊戌变法是一次具有爱国救亡意义的变法维新运动，是中国近代史上一次重要的政治改革，也是一次思想启蒙运动。戊戌变法促进了思想解放，并且对思想文化的发展和中国近代社会的进步起到重要的推动作用。

图 21-1　中南海

图 21-2　中南海中的瀛台

21.2.2 中华人民共和国成立

随着戊戌变法的失败与历史的变迁，1912 年中华民国成立，清帝逊位后，中南海被袁世凯占用，成为北洋政府总统府。袁世凯称帝时期，中南海曾改名"新华宫"，同时宝月楼拆外墙，更名为新华门（图 21-3）。

中南海先后被用作北洋政府的总统和总理办公地，以及张作霖"帅府"（图 21-4）。国民政府迁都南京后，中南海作为公园对民众开放。

图 21-3　新华门

1949年后，中南海成为中国共产党中央委员会和中华人民共和国国务院的驻地和部分国家领导人居住的地方。

1949年9月21日—30日，中国人民政治协商会议第一届全体会议召开，地点设在中南海富丽堂皇的怀仁堂。怀仁堂原本是清宫御苑一个巨大的四合院，中央是天井。建筑学家梁思成奉毛泽东之命，在天井上加盖了一个可容纳千人的礼堂，与原有宫殿古建筑浑然一体，天衣无缝，既省时省钱，又具有民族特色，表现了艺术家的匠心独运。怀仁堂成为党的领导人与政协代表研讨开国大计的殿堂（图21-5）。

图21-4　张作霖"帅府"

图21-5　怀仁堂

1949年10月1日下午2时，毛泽东在中南海主持召开中央人民政府委员会第一次会议，中央人民政府主席、副主席、委员宣布就职，中央人民政府成立。

Project 1　From "walls fall down, the house does not collapse"

1.1　Theoretical guidance

1.1.1　The three dimensions of "walls fall down, the house does not collapse" –building materials

The most graphic description of traditional Chinese architecture is that "walls fall down, the house does not collapse", the secret of which is hidden in the building materials and structure. The three dimensions of a building are the materials, the four beams and eight columns and the mortise and tenon construction (Figure 1–1).

Figure 1–1　Wood

Characteristics of wood:

(1) Wood is light in weight and flexible, while wooden structures are light in self-weight.

(2) Wood elements are easy to transport, assemble and disassemble and can be used many times.

(3) Wood is also fire and corrosion resistant, which is why it is often used as a material for building houses.

1.1.2 "Four beams and eight columns"

Regardless of the scale of the building, the outline of an ancient Chinese building consists of a pedestal, a wooden part of beams and columns and a roof (Figure 1-2).

Figure 1-2 Ancient building construction

Most of the ancient buildings were built with a three-room layout, and because of the large openings, the middle to add two beams, plus the front beam and the back beam together are four beams, and each beam has a column at each end to play the role of support. So there are eight columns. By these four beams and eight columns support the whole building. The walls are used to separate the space, not mainly for support. In this way, there are four beams and eight pillars, representing the meaning of the main structure (Figure 1-3).

Figure 1-3 Internal structure of ancient buildings ("four beams and eight columns")

1. Arch

The arch, as a wooden structure part, its role is to extend the cantilever to support the weight of the eaves part (Figure 1-4).

2. Opening

The open space is the distance between two adjacent eaves columns on the front, also known as the "face width". Eaves column is the area covered by the four figures above, that is, the outermost column of the house, the distance between the two eaves columns (Figure 1-5). The openings in ancient buildings are generally odd openings, such as a room, three rooms, five rooms, seven rooms, nine rooms.

Figure 1-4 Arch Figure

Which has the most on the open room is the Forbidden City in the Palace of Taihe Hall has 11 open rooms, is the highest architectural system in ancient buildings. Folk buildings are generally used three rooms or five rooms, because the folk building scale is smaller.

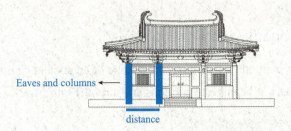

Figure 1-5   Diagram of ancient building openings and eaves and columns

1.1.3 Mortise and tenon construction

Mortise and tenon construction is the soul of ancient architecture (Figure 1-6). Mortise and tenon is a unique way of connecting ancient Chinese wooden elements by using concave and convex interlocking. In mortise and tenon construction, the tenon is the protruding part of a wooden member, also called the mortise. The part of the tenon that extends into the mortise is called the tongue, and the rest of the tenon is called the shoulder. The mortise is the recessed part, also known as the mortise and tenon slot. The tenon and mortise bite into each other to create a supporting force of "1+1>2".

Figure 1-6   Mortise and tenon construction

Mortise and tenon was widely used in architecture, furniture and other wooden instruments, and mortise and tenon techniques reached the highest level in the Ming and Qing dynasties.

The mortise and tenon structure is unsurpassed in the history of construction, without a nail or glue, the shape of the wood is repaired to fit each other respectively, and the shape between the two can be combined together and bite each other, which has achieved the effect of stability.

The mortise and tenon structure is also a basic skill that ancient carpenters must have, and the level of craftsmanship can be clearly reflected through the structure of mortise and tenon. Lu Ban, who is respected as the originator of Chinese carpenters, brought the mortise and tenon structure to the extreme, and the spirit of great craftsmanship permeates between mortise and tenon.

The combination of mortise and tenon is the key to earthquake resistance. This structure is very similar to the modern beam and column frame structure, the nodes of the frame are not rigid, which ensures the coordination of the building's rigidity. In the event of a strong earthquake, the mortise and tenon combination of space structure will be "loose" but not "loose", it allows the structure of the house to produce a certain amount of deformation, and thus consume the energy transmitted by the earthquake, so that the earthquake load of the entire house is greatly reduced, which plays a role in earthquake resistance.

The ancient Chinese craftsmen created the exquisite arch and mortise and tenon structure, which is full of wisdom and design, reflecting the Eastern philosophical way of thinking-to overcome rigidity with flexibility.

1.2 Interpretation of the spirit of traditional Chinese culture

The ancient Chinese architecture of "walls fall down, the house does not collapse" can be considered a miracle in the history of architecture, whether it is the love of the ancients for wooden materials, or the four beams and eight pillars that support the entire building, and then the precise mortise and tenon structure, each of which exudes the wisdom of the ancients and contains the profundity of Chinese culture, the language of construction of those delicate structures speaks from the the language of construction speaks to the distinct humanist spirit of ancient civilizations.

1.3 Project innovation

Talk about the essence of ancient Chinese architecture and your understanding and thoughts on the wisdom of the ancients in building with the principle of "walls fall down, the house does not collapse" in the course content.

Project 2　Table base and ground

2.1　Theoretical guidance

2.1.1　Basic structure of ancient buildings

From the shape of an ancient building is clearly divided into three parts: the foundation, the body, and the roof.

Yu Hao, a famous craftsman in the Northern Song Dynasty, called the "three points" in *the wood* and pointed out that "where the house has three points, from the beam above the upper points, above the ground for the middle points, the steps for the lower points".

According to the ancient building houses these three parts of the composition of the function, the ancient building structure can be divided into the following parts. The upper part—the roof; the middle part—the body of the house wooden frame, arch, wall; the lower part—the foundation, foundation, ground (Figure 2–1).

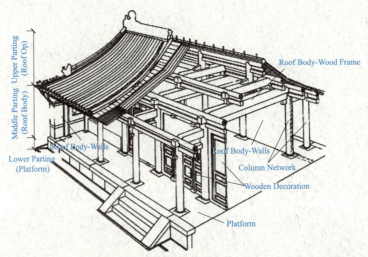

Figure 2–1　Composition function of three parts of ancient buildings

2.1.2 The role of table base and pedestal

The pedestal is the base of the building above the ground, a prominent platform made of masonry to support the building, and make it moisture-proof and anti-corrosion. It is used to support the building and make it moisture-proof and anti-corrosive, and to make up for the lack of tall and majestic monolithic buildings of Chinese ancient architecture (Figure 2–2).

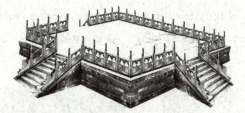

Figure 2–2   Table base

2.1.3 Form of pedestal

According to the use function and shape, the table base is broadly divided into two categories: ordinary table base and Sumeru base.

1. Ordinary table base

From the early relics, the earliest foundation was made of rammed earth, which is presumed to have appeared at least in the Shang Dynasty (Figure 2–3). The foundation made of rammed earth is also called rammed earth platform. Rammed earth is to compact the clay, reduce the gap between the clay, increase the density of the clay mixture.

Han Dynasty has begun to wrap masonry, to replace the rammed earth, so called the plinth （sǎng）pier (Figure 2–4).

There is also a kind of higher pedestal, higher than the ordinary pedestal, often built on the top side of the pedestal Hanbai jade railings, used for large-style buildings or secondary buildings in the palace building (Figure 2–5).

Figure 2–3   Shang Dynasty rammed earth platform

Figure 2–4   Han Dynasty plinth （sǎng）pier

2. The Sumeru pedestal

The word "Sumeru" was originally translated from Sanskrit and is said to have been a famous mountain in ancient Indian mythology. The word "Sumeru" means "wonderful height", "wonderful light" "good accumulation" and so on. This is why Mount Sumeru is also sometimes translated as "high mountain", meaning a mountain of treasures. The name Sumeru is also known as "Vajra Seat" or "Sumeru Altar", which later came to refer to the base of an architectural decoration.

In feudal society it was only used in high-ranking palaces, monasteries, Taoist temples, etc., as well as on some monumental buildings. As a result, it was deeply marked by hierarchy (Figure 2–6—Figure 2–9).

Figure 2–5   Higher pedestal

Figure 2–6   The Sumeru pedestal

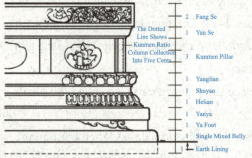

Figure 2–7   Song-style Sumeru

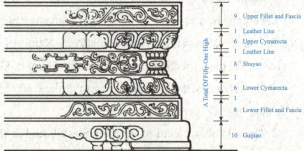

Figure 2–8   Qing-style Sumeru

The dragon's head in the Sumeru Throne in the Palace of the Forbidden City is a drainage system that displays the spectacle of a "thousand dragons spitting out water" during heavy rains (Figure 2–10).

The structure and style of the Sumeru Throne:

The Qing-style Sumeru pedestal is generally made of stone, and from the top to the bottom are: the upper square (fāng), the upper lord (xiāo), the girdle, the lower lord, the lower square and the foot of the gui, which can also be divided into carvings of the girdle alone, carvings of the girdle and the upper square, and all carvings of the Sumeru pedestal (Figure 2–11, Figure 2–12).

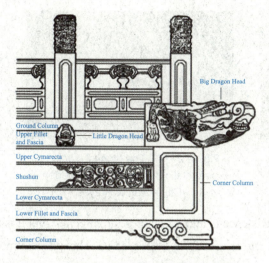

Figure 2-9　Qing style with tapering whisker seat

Figure 2-10　The "Thousand Dragons Spitting Water" Spectacle in the Forbidden City

Figure 2-11　Sumeru Structure

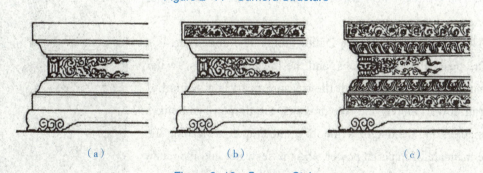

Figure 2-12　Sumeru Style

(a) Xumizuo corved at the waist;　(b) Xumizuo carved on the waist and upper brace;　(c) All carved Xumizuo

3. The highest level pedestal

The highest level of pedestal by several Sumeru seat stacked, so that the building appears more magnificent and tall, often used in the highest level of construction, such as the Forbidden City three halls and the Temple of Heaven praying for the Temple, that is, towering on the highest level of pedestal (Figure 2-13, Figure 2-14).

Figure 2-13　The highest-ranking pedestal in Palace's Taihe Hall

Figure 2-14　The highest-ranking pedestal in the Imperial Forbidden City's Hall of Prayer

Not only does it have a long history of existence in ancient Chinese architecture, but it is also very widespread and can be found in palaces and houses from the top down. This phenomenon is naturally linked to the various benefits of using a pedestal.

The foundation has a structural load-bearing function. They are better able to carry the weight of the upper part of the building and prevent uneven settlement from occurring. As the construction of a foundation was a very demanding and technically complex task in an era of low productivity, the presence or absence of a foundation and its height naturally became a marker of status.

4. The Danbi Stone

It is also known as the Stepping Stone, is a large rectangular stone inlaid in the middle of the steps in front of the ancient palace, usually a whole stone, ordinary people can not walk on the Danbi Stone, but only from its sides, is a symbol of imperial power.

5. The symbolic meaning of the relief patterns of the Dangbi Stone

The relief patterns on the Danbi Stone. The upper part is a treasure pearl and two dragons, and the lower part has seawater and mountain stones, of which the treasure pearl is a symbol of imperial power; among the two dragons, the dragon symbolizes the emperor, and the dragon symbolizes the emperor, meaning the divine mandate of imperial power; and the seawater and mountain stone patterns are symbols of the eternal stability of the rivers and mountains (Figure 2-15).

Figure 2-15　The Dangling Stone Relief symbolism illustration

2.1.4　Stomp and the expression of stomp

Stomp, a step in ancient Chinese architecture, is generally made of brick or stone and placed between the pedestal and the outdoor ground. The forms of stomp include: pendant stomp, ruyi stomp, Concretionary stones and imperial road stomp (Figure 2-16).

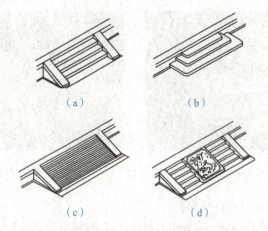

Figure 2-16　Stomp Style
（a）Drooping Belt Stone；　（b）Ruyi Stone；　（c）Sidestep Stone；　（d）Yulu Stone

1. Draping Steps

The draped stomp is a very neat step, with a slanted stone on each side, called "draped stone" (Figure 2-17).

There is also a more advanced type of draped tread called a triple step (Figure 2-18). In ancient times, the high platform building is an important royal building, in order to facilitate up and down the high platform, also to build steps, the stone steps used in this high platform is the "steps". In order to distinguish between the superior and the inferior, the stone steps on the upper and lower platform were arranged in three ways, left, centre and right, called the "three outstretched steps". The centre of the steps is wider, with a carved stone with a narrow step on both sides; the left and right steps are equally wide, but are narrower than the middle steps for officials to walk.

Figure 2-17　Draping Steps

Figure 2-18　Three Exits

2. Ruyi stomp

Ruyi stomp is a step that reduces the length and width of each slab, so that people can go up and down from different directions. Other maintenance facilities are provided on each side (Figure 2-19, Figure 2-20).

Figure 2-19　Ruyi stomp

Figure 2-20　Natural ruyi stomp

3. Concretionary stones （jiāng cā）

Concretionary stone is a sloping surface made of slabs or bricks instead of steps. To prevent slippage, horizontal mouldings are cut into the sloping stone surface (Figure 2-21).

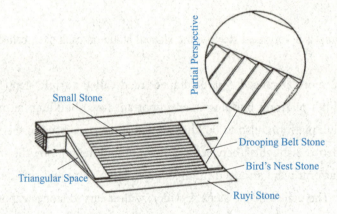
Figure 2-21　Concretionary stones （jiāng cā）

4. Imperial Stamp

The imperial stomp, generally used in palace and temple buildings, was originally a Chinese palace building form, a ramp between the foundation and the floor and the steps on either side of the palace's central axis; in feudal times only the emperor could use it. In the middle of the path, there is a stone staircase with balustrades on both sides (Figure 2-22).

5. The Dragon's Tail Path

"The Dragon's Tail Path" originated in the Warring States period as a high platform building, and as the height of the building was lowered, the path was high in the front and low in the back, collapsing to the ground. It looks like the tail of a dragon hanging down, hence the name The Dragon's Tail Road (Figure 2-23).

Figure 2-22　Imperial Stamp

Figure 2-23　The Dragon's Tail Path

2.1.5　Ancient building grounds

The grounds of ancient buildings include canopies and loose water.

1. Canopied roads

A canopied path is a path made of masonry (Figure 2-24).

Depending on the fineness of the masonry, the floor can be divided into a rough plastered floor and a fine plastered floor. The word "plastered" here means painted and decorated.

The brown plastered floor means that the brick material does not need to be cut and polished, the joints of the floor tiles are wider, and the difference in height between the brick and the brick adjacent to it and the flatness of the floor are not as elaborate as the fine plastered floor (Figure 2-25).

Figure 2-24　Canopied roads

The fine plastered floor means that the brick material should be processed by cutting and grinding, after which the brick specifications are uniform and accurate, the angles are complete and straight, and the surface is smooth and clean. The most elaborate of these was the "gold brick plastering", which was generally only used in the most important buildings such as palaces (Figure 2-26).

Figure 2-25　The brown plastered floor

Figure 2-26　The fine plastered floor

2. Scattered water

The water scattering refers to the slope of the water scattering around the external walls of the house, on the outdoor floor, with a certain slope. The function of the scattered water is to quickly drain away the rainwater from the vicinity of the house, to prevent it from washing away or infiltrating the foundations and to prevent the foundations from sinking, in order to ensure the consolidation and durability of the house (Figure 2-27).

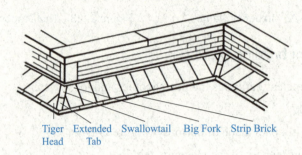

Tiger Head　Extended Tab　Swallowtail　Big Fork　Strip Brick

Figure 2-27　Scattered water in ancient buildings

2.2　Interpreting the spirit of traditional Chinese culture

Through the understanding of the foundations and floors of ancient buildings, we have a deep sense of the wisdom of the ancient working people, whose long history has created a splendid Chinese culture, of which ancient architecture is an important part. The basic threads of the rational spirit of the planes are still present in the clay foundations and the footsteps, still translating spatial awareness into a temporal process, and still expressing the realities of life in the world.

2.3　Project innovation

Write a reflection on "The Wisdom of the Chinese People in Ancient Chinese Architecture" through the study of "Foundations and Grounds" in any form.

Project 3　Wall construction

3.1　Theoretical guidance

3.1.1　Types of ancient architectural walls

The Great Wall of China is a representative of ancient architectural walls, with a large number of cities, barriers, pavilions and markers combined with a defense system, while the walls were built with brick, stone and brick mixed masonry, so let's explore together the wisdom of the ancients hidden in the walls.

The wall belongs to the middle part of ancient architecture, according to the position of the wall in the building is divided into types such as mountain wall, eave wall, threshold wall, fan wall (after the gold wall), partition wall.

1. Mountain wall

Mountain wall is generally known as the outer horizontal wall, refers to the wall along the short axis of the building direction-arrangement called horizontal wall, the horizontal outer wall at both ends of the building is generally known as the mountain wall. Ancient buildings generally have a mountain wall, its role is mainly separated from the neighbor's residence and fire prevention (Figure 3–1).

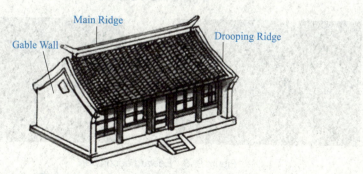

Figure 3–1　Ancient building mountain wall

There are three types of mountain walls, one is herringbone, this shape is more simple and practical, and the cost of construction is not high, the folk mostly use this kind of mountain walls (Figure 3-2).

Pot ear-shaped, beautiful lines, large changes, in fact, it is built after the ancient official hat shape, to take the meaning of the future, the general people's homes are often used (Figure 3-3).

Wave-shaped, the shape of the undulation, pay attention to symmetry, undulation more than three levels, the actual pot ear wall is a deformation, more like the ancient official hat, the people basically do not use (Figure 3-4).

 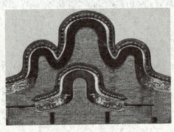 

Figure 3-2   Herringbone Mountain Wall

Figure 3-3   Pot ear-shaped mountain

Figure 3-4   Wave-shaped mountain

2. Edward（chí）

Because of its special position, from a distance, it looks like the neck of a house, so the owner of the house will take great pains to decorate it. The subtlety of the carving can be seen in every trace of it (Figure 3-5, Figure 3-6).

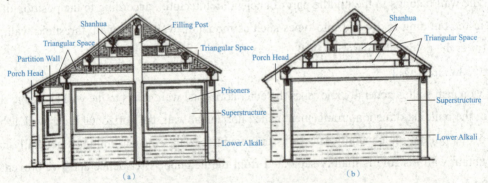

Figure 3-5   Edward

Figure 3-6   Edward decoration

3. Gable Wall

The wall between the gable and the eaves is called "the gable wall", which is the external longitudinal wall of the building, customarily called the gable wall or external longitudinal wall. The vertical wall is called "the front gable wall" in the front gable of the building and "the rear gable wall" in the rear gable of the building.

4. Sill wall

The sill wall is the wall under the wooden decorative couch board at the front or rear eaves of a building (Figure 3-7).

5. Horizontal partition wall

Horizontal partition wall, also known as horizontal partition walls within the gold, mainly refer to the walls between the front and rear gables and the gold pillars (Figure 3-8).

The outermost part of the building is called the eaves column. The one located on the central axis of the building is called the mountain column, and the one located between the eaves column and the mountain column is called the gold column (Figure 3-9). In this context, as long as the pillars are not located in the central axis of the building and the outermost position are called gold pillars, the wall between the gold pillars is called the horizontal partition wall.

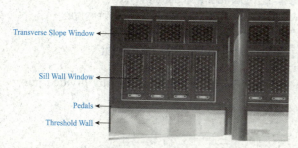

Figure 3-7   Sill wall

Figure 3-8   Horizontal partition wall

6. Partition Wall

The partition wall, also known as the shelf mountain or sandwich mountain, is built between the front and rear gable columns and the inner wall parallel to the mountain wall (Figure 3-10).

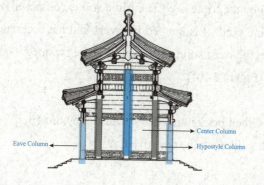

Figure 3-9   The name of the ancient building pillar

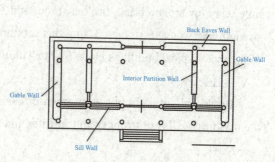

Figure 3-10   Partition Wall

7. Corridor wall

Corridor wall, also called "corridor heart wall", is the wall between the eaves and gold pillars on the inside of the wall of the building (Figure 3-11). The center of the corridor wall can be decorated in a variety of forms, such as the plain heart, paintings, and even carvings; the contents can be geometric patterns, auspicious patterns such as the Wan Zi pattern, or flowers, birds and animals (Figure 3-12).

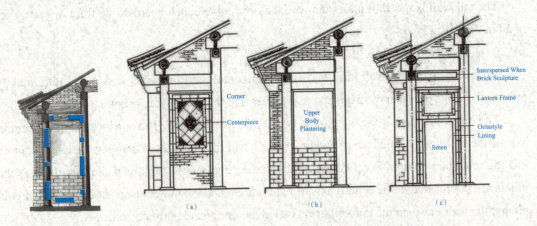

Figure 3-11 Corridor wall

Figure 3-12 Corridor wall decoration
(a) Gallery wall decoration; (b) Plain wall type; (c) Door copper type

3.1.2 Classification of ancient building wall forms by material

According to the classification of ancient building walls by material, mainly divided into earth walls, brick walls, stone walls.

1. Earth wall

(1) Rammed earth wall. Rammed earth wall is one of the oldest forms of walls in China. Earthen walls have good heat insulation, sound insulation, and a certain bearing capacity, and can be locally sourced materials, construction is also very simple, but vulnerable to natural erosion, especially after flooding the strength of the wall is greatly reduced, so the ancient construction of walls pay close attention to site selection and drainage (Figure 3-13).

(2) Sanhe earth wall. The wall is rammed by the triple clay. The triple soil is composed of lime, clay (or broken brick, broken stone) and fine sand. It is tamped in layers and has a certain strength and water resistance. Using this mixture, the most solid earth walls can be rammed. One of the most typical buildings is the Fujian Tulou (Figure 3-14).

2. Brick wall

Brick wall, a wall built with green bricks. Green bricks are hard because they are fired, so the masonry walls are stronger and sturdier and less likely to be destroyed.

Figure 3-13　Rammed earth wall　　Figure 3-14　Fujian Tulou sanhe earth wall

The use of brick was recorded in China at an early stage, but initially it was mostly used in the construction of tombs and towers, and less in ordinary residential buildings. It was not until the Ming Dynasty that real brick walls became more common (Figure 3-15).

Figure 3-15　Brick wall

(1) Clear water wall. Where the surface of the wall without painting, without veneer material brick wall, are called "clear water wall". Clear water wall brick joints with sand or mortar hook fill, in order to prevent the invasion of rain and wind. The appearance of the wall after hooking also looks more neat, although not gorgeous, but looks simple and clean (Figure 3-16).

(2) Hollow wall. Hollow brick wall, with obvious material-saving characteristics, is very economical-but its stability is not so bad sometimes such a hollow brick wall can also be used as a load wall -and also has good sound insulation performance, that is, economic and practical (Figure 3-17).

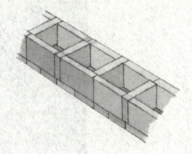

Figure 3-16　Clear water wall　　　　　Figure 3-17　Hollow wall

① Leaky brick wall. Leaky brick wall is a kind of fancy brick wall, that is, at the hole of the wall with bricks to build a diamond or make bamboo knots and other carvings (Figure 3-18).

Figure 3-18   Hollow wall

②Leaky brick wall (Figure 3-19). Window wall is with leaky windows, which is common in garden buildings. First of all, the wall plays the role of isolating scenic spots, but the artistic conception of gardens requires that it cannot be completely closed but separated continuously. Therefore, this wall with leaky windows is called window wall.

Figure 3-19   Leaky brick wall

③Brick flower wall (Figure 3-20). Tiled wall refers to the pattern with rhythmic beauty formed on the wall surface through the uneven change of brick body. The practice is diverse.

Figure 3-20   Brick flower wall

3. Stone Wall

A wall made of stone is called "a stone wall". Stone is a natural material and is very malleable. One of the most typical stone walls is the tiger skin stone wall.

The tiger skin stone wall is a kind of stone wall used in ancient architecture, built with irregularly shaped rough stones, with gray joints between the rough stones, and the gray joints match the contours of the stones (Figure 3-21).

From the perception, especially from a distance, the tone, texture of the pattern is somewhat similar to the tiger skin, so the master craftsmen have agreed to call this kind of wall like tiger skin stone wall, continues to this day. The tiger skin stone wall is rustic in appearance and is mostly used in temples and mansions.

Figure 3-21　The tiger skin stone wall

3.1.3　The role of the walls of ancient buildings

The role of the walls of ancient buildings is divided into five points.

1. Load-bearing role

Load-bearing role, mainly applied to brick and wood mixed structure of the house. As the eaves wall without wooden columns roof frame load by the front wood frame and the eaves wall together.

2. Enclosing and separating interior space

The wall is usually divided into the interior or left, center, right three parts, forming a "one place, two inside" the basic pattern.

3. Fire prevention role

Ancient building wall materials are mostly brick, adobe, stone, etc., its fire performance is much higher than wood.

4. Security defense role

Ancient building walls to expand the construction can form walls, walls, the Great Wall, etc., is a concentrated embodiment of defensive structures, in ancient warfare, playing an important defensive role.

5. Decorative role

Ancient buildings in the wall masonry has a strict process requirements, different walls can show different artistic effects. For example, the leaky brick wall, leaky window wall and brick flower wall.

3.2 Interpretation of the spirit of traditional Chinese culture

The wall is like a mirror through which people of different times reflect on themselves, and the wall is like a warrior, standing majestically, guarding the home. A wall with a different landscape, a wall full of scenery, a wall outside a hundred states of life. Brick flower wall masonry form of concave and convex changes, rhythmic undulation formed a different pattern change effect, to achieve a different sensory experience. The exquisite skills and wisdom of the ancients reveal the immortal legend in the art of construction from brick to brick.

3.3 Project innovation

Write a reflection on "the wisdom of the nation hidden in ancient Chinese architecture" through the study of "wall construction", in any form.

Project 4　Wooden frame construction

4.1　Theoretical guidance

4.1.1　Beam-bearing wooden frames

Traditional wood frame structure is mainly divided into three types of forms: beam-bearing wood frame, a kind of wood frame through the bucket and well-dry wood frame. The beamed timber frame is mostly used for important wooden buildings, the bucket frame is generally used for folk buildings, and the well-dry frame is mostly used in some special areas, such as this area has a lot of wood and no other materials.

Beam-bearing frame is the main form of wood frame in ancient Chinese architecture. This type of frame is characterized by several layers of stacked beams along the depth of the house, the beams are shortened layer by layer, short columns or blocks are padded between the layers, and small columns are set up in the middle of the uppermost layer of beams to form a triangular roof frame. Because its beams are stacked layer by layer, so it is also called "stacked sorghum type frame" (Figure 4–1, Figure 4–2).

Characteristics of beam-bearing wooden frames:

Beam-bearing wooden frame structure is complex, requiring detailed processing, but strong and solid, durable, and the interior has a large use of space, but also to produce a magnificent momentum and beautiful shape (Figure 4–3).

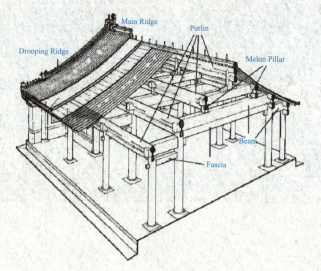
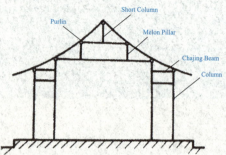

Figure 4-1   Form of beam-bearing wooden frames

Figure 4-2   Diagram of the beam-bearing timber frame

Figure 4-3   Beam-bearing wooden frames structure

4.1.2   Bucket-type wooden frames

The bucket-type wooden frame is a kind of simple and light frame. The frame is also made up of columns along the depth of the house, but the columns are more closely spaced, and the columns bear the weight of the purlins directly without overhead lifting beams. This kind of frame is commonly used in southern China (Figure 4-4).

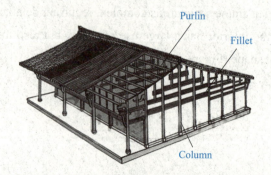

Figure 4-4   Form of bucket-type wooden frames

4.1.3 Well-dry wooden frames

Well dry structure is a kind of house structure without columns and beams. This structure is made of round wood or rectangular or hexagonal wood stacked in parallel layers upwards, and the ends of the wood are crossed and nibbled at the corners to form the four walls of the house, shaped like the wooden fence on the ancient well, and then short columns are set up on the walls on the left and right sides to bear the ridge purlins to form the house (Figure 4-5, Figure 4-6).

Figure 4-5 Form of well-dry wooden frames

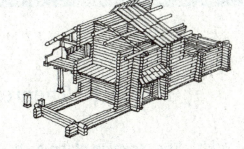
Figure 4-6 Structure of well-dry wooden frames

4.2 Interpretation of the spirit of traditional Chinese culture

Chinese ancient architecture with wooden structure as the main body is unique in the world's architectural forest with distinctive style. It is built with wooden structure as the bone, brick and stone as the body, knotted tile as the cover, oil painting as the clothes, carefully designed by generations of skilled craftsmen, clever construction, dedicated decoration, and put their efforts and wisdom into building, reflecting the unique artistic charm of classical Oriental architecture and the history, artistry and science of Chinese ancient architecture wooden structure.

4.3 Project innovation

Draw a mind map of the structure of ancient buildings from foundations and floors, wall construction, to timber frame construction, in any form you wish, taking into account what you have learned in the previous lessons.

Project 5 Arch structure

5.1 Theoretical guidance

5.1.1 The meaning of the arch

The arch is a wooden element used to support the eaves of a house (Figure 5–1).

Figure 5–1 Form of arch

5.1.2 The role of arch

1. Increase the distance

The arch can pick out the purlins on the outermost eaves by a certain distance, making the eaves of the building more far-reaching and the shape more beautiful and spectacular.

2. Load function

The bucket is located between the column and the beam, the load from the roof and the upper frame, to pass through the bucket arch to the column, and then from the column to the foundation, so it plays the role of bearing the top and passing the load (Figure 5–2).

3. Decorative function

Arch construction is delicate, beautiful shape, such as bonsai, like a flower basket, and is a very good decorative elements. The arch gradually became decorative, because later the brick wall, the eaves became much closer, so the arch role is reduced, but also to distinguish between the signs of architectural class. The more noble the building arch, the more complex and complicated (Figure 5–3).

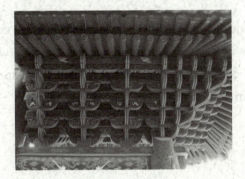
Figure 5–2 Load function of arch

Figure 5–3 Decorative function of arch

4. Anti-earthquake function

The combination of mortise and tenon is the key to earthquake resistance, and the arch holds the weight of the eaves evenly, playing a balancing and stabilizing role. The key to the "wall not collapsing" of ancient Chinese buildings is the mortise and tenon structure of the wooden components (Figure 5–4).

Figure 5–4 Yingxian wooden pagoda

5.2 Interpretation of the spirit of traditional Chinese culture

In traditional ancient architecture, ancient craftsmen then used short pieces of wood protruding from columns and beams, arches supporting buckets and buckets supporting arches, layered on top of each other, to form the most beautiful elements of ancient Chinese architecture.

From the royal family's exclusive use to the people, the arch is a historical record of Chinese architecture. When you are under the arch, there is always a stirring feeling of traveling through the past and the present, being under the same blue sky as the ancient people a thousand years ago.

5.3　Project innovation

Combine what you have learned in the previous courses, from "walls fall down, the house does not collapse" to the foundations and floors of ancient buildings, wall construction, wood frame construction, and arch construction, talk about your impressions and feelings about ancient Chinese architecture, either in video form or in written form.

Project 6　Roof construction

6.1　Theoretical guidance

6.1.1　Roof form of ancient buildings

Roofs of ancient buildings can be divided into: hard hill roof, overhanging roof, hipped roof and hyperset from the form.

1. Hard hill roof

The hard hill roof has two characteristics: the two-sided slope of the roof, the eaves against the wall of the hill do not pick up (Figure 6–1).

2. Overhanging roof

The overhanging roof is also a two-sided slope, but the biggest difference between it and the hard roof is that the eaves on both sides of the mountain wall are the pick out, and the pick out beyond the mountain wall is the overhanging mountain (Figure 6–2).

Figure 6–1　Hard hill roof

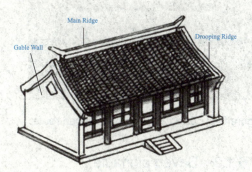
Figure 6–2　Overhanging roof

3. Hipped roof

The most obvious sign of a hipped roof is that it has a four-sided slope and is the highest specification of the four types of roof systems (Figure 6–3).

4. Hyperset roof

The hipped roof is equivalent to the feeling of cutting two knives on the hyperset roof, that is, cutting two knives on the roof above the wall to form a half-slope, and at the same time, there are four additional bumpy ridges (Figure 6–4).

Figure 6–3　Hipped roof　　　　　　　　Figure 6–4　Hyperset roof

Among the roof grades of ancient buildings, the highest grade is the hipped roof, followed by the hyperset roof, then the overhanging roof, and finally the hard roof.

6.1.2　Overlapping eaves

A two-story roof is called overlapping eaves in architectural language, and overlapping eaves are generally divided into overlapping eaves hipped roofs and overlapping eaves hyperset roofs (Figure 6–5, Figure 6–6).

Figure 6–5　Overlapping eaves hipped roofs　　　Figure 6–6　Overlapping eaves hyperset roofs.

6.1.3　Saved pinnacle

The roof is topped by a lovely pagoda-like element called a paulownia, and a roof made of paulownia is called a saved pinnacle (Figure 6–7, Figure 6–8).

6.1.4 The names of the ridges on the roof

The bump ridge is called the ridge with the front eaves of the house at an angle of 45 degrees, and the top one is called the main ridge; the one with the main ridge at 90 degrees is called the vertical ridge; the mountain flower is the triangle connected with the vertical ridge (Figure 6–9).

Figure 6–7　Paulownia　　　　　Figure 6–8　Saved pinnacle

Figure 6–9　The names of the ridges on the roof

6.1.5 Wan Zi Ding

The building of Wan Zi ding is a palace hidden in the middle of Yuanmingyuan, which means peace and harmony for all parties (Figure 6–10, Figure 6–11).

 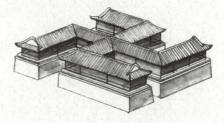

Figure 6–10　The ruins of the Wan Zi ding Palace in Yuanmingyuan　　　　Figure 6–11　Wan Zi ding Palace

6.2 Interpretation of the spirit of traditional Chinese culture

How many stories have been hidden in the roofs of China for thousands of years? They are silent in history, proud of the world, weathered and baptized, but still standing firm, holding up a far-reaching sky. Standing on the rooftop to see China is to open a history book carved in stone and encounter the brightest scenery in a thousand years of wind and rain. Hard roofs, overhanging roofs, hipped roofs, and hipped roofs are all writing beautiful chapters on Chinese ancient architecture.

6.3 Project innovation

Combined with the content of the previous courses, from the ancient building's foundation and ground, wall construction, wood frame construction, arch construction and today's study of roof construction, we already have a certain understanding of the entire ancient building house structure. The structure of these ancient buildings is organized into a mind map of ancient architecture according to the three sections of ancient architecture.

Project 7　Wood decoration

7.1　Theoretical guidance

7.1.1　Interior eaves decoration of ancient buildings

In the construction of ancient Chinese buildings with wooden structure system as the main body, the main wooden structure of the work, such as columns, beams, square, purlins, called large woodwork. Corresponding to this, building doors and windows, ceilings, indoor partitions and other wood structure part of the work called small woodwork. This woodworking division of labor is clearly recorded in the Song Dynasty *Yingzao Fashi*, has been used to this day. Small woodworking to the Qing Dynasty is also known as wood decoration, and subdivided into the exterior eaves decoration and interior eaves decoration. The exterior eaves decoration includes doors, windows and exterior eaves column decoration; interior eaves decoration includes flower cover class, ceiling algae well and wooden stairs

1. Exterior eaves decoration-door

Door is one of the most important components of ancient Chinese architecture, it is not only a channel for people to enter and exit, but also a symbol of architectural rank, indicating the status, wealth, cultural character.

（1）Solid door. Solid door commonly used in palaces, royal palaces and other higher-grade buildings, the door is often decorated with gold nails, as a symbol of rank and status (Figure 7-1).

（2）Chessboard door. Checkerboard door because its door looks like a chessboard, so called "checkerboard door" (Figure 7-2).

（3）House-style gates. The main form of the gate is the house, which appears as a single house, both as a door and as a house. It was widely used from the emperor's palace down to the homes of ordinary people (Figure 7-3).

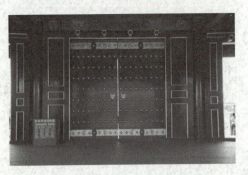

Figure 7-1　Solid door

Figure 7-2　Chessboard door

（4）Door ring.The "door ring" is also called "store head", which is a practical object for opening and closing the door and knocking (Figure 7-4).

Figure 7-3　House-style gates

Figure 7-4　Door ring

2. Exterior eaves decoration-window

The window is like the eye of the building, showing the building's charm through its deep eyes and blending in with the architectural style. At the same time, the windows of ancient Chinese architecture are not only practical but also aesthetically pleasing, overflowing with the colours of oriental aesthetics.

（1）Straight-panelled windows. The straight latticed window is the simplest style of window, with the lattices arranged vertically like a fence (Figure 7-5).

（2）Broken Pane Window. A broken pane window is characterized by the word "broken", as its pane is made by breaking a square section of wood diagonally along the diagonal, i.e. the tip of the triangular section faces outwards and the flat side faces inwards, so that it can be easily papered inside the window to keep out wind and sand, cold air, etc (Figure 7-6).

Figure 7-5　Straight-panelled windows

Figure 7-6　Broken Pane Window

（3）Sill window. A sill window is an advanced form of window, a partition window with a pivot on the top and bottom of the sash that can be opened and closed inward and outward (Figure 7–7).

（4）Branch window. The branch window is a kind of window that can be pushed up and taken down, and has been commonly used in ordinary houses since the Ming and Qing dynasties. The window is generally divided into two sections, the upper end can be pushed up to support, the lower end can be taken off, which is the origin of the name of the window, but also the biggest difference between it and the sill window (Figure 7–8).

Figure 7–7　Sill window　　　　　　Figure 7–8　Branch window

（5）Leaky windows. Leaky windows are a type of windows that are more free in form, but cannot be opened. Through the window, we can see the scenery on the other side, which seems to be separated and hidden. Later, most of the windows were developed with colorful patterns built in, which are beautiful spots in themselves (Figure 7–9).

（6）Empty windows. The difference between an empty window and a leaky window is that an empty window has only a window hole but no window pane. The setting of empty windows can make several spaces interspersed and penetrate each other, integrating the internal and external scenery, and increasing the depth of field and expanding space (Figure 7–10).

（7）Assorted windows. The charm of assorted windows not only comes from its artistic shape, but also from the color and decoration of the window cover, rich in brick carving themes, including natural flowers and plants, auspicious patterns, etc. The regular and orderly arrangement of patterns in the window center part of the lattice not only gives people a beautiful visual effect, but also gives people a beautiful psychological feeling from the rich moral it contains (Figure 7–11).

Figure 7–9　Leaky windows　　　　　　Figure 7–10　Empty windows

Figure 7-11 Assorted windows

3. External eaves column decoration

The decoration of the exterior eaves column includes lintel, balustrade, and finial.

(1) Lintel. Lintel is a decoration with both decorative and practical functions installed between the eaves columns of the building (Figure 7-12, Figure 7-13). Depending on the location, it is divided into hanging lintels and stool lintels. Hanging lintel is installed under the eaves square, which has the function of enriching and decorating the building façade; sitting lintel is installed between the columns under the eaves, which has the function of enriching the façade, but also allows people to sit and rest.

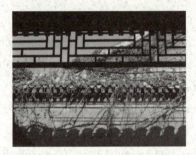

Figure 7-12 Hanging (inverted) lintel

Figure 7-13 Stool lintel

(2) Railings. Railings are initially used as a shield, and then gradually developed and changed, rich in style, carved exquisite, became an important decorative settings. And in the garden, the railing plays a role in the role of partition and scenery, function like a leaky window, and the image of class flower wall. The column in the railing is called the column, the middle part of the two columns is called the railing plate (Figure 7-14).

Figure 7-14 Watchposts and balustrades in railings

(3) Sparrows. Originally placed at the upper end of a column to bear the upper pressure together with the column, it was positioned at the junction of the beam and the column or the square and the column. As a decorative element of the eaves of ancient buildings, it is the ultimate carving technique and carries a rich cultural connotation (Figure 7–15, Figure 7–16).

Figure 7–15 Sparrows (1)

Figure 7–16 Sparrows (2)

7.1.2 Interior eaves decoration of ancient buildings

1. Interior eaves decoration—flower covers

The flower cover is mostly used for interior decoration and partitioning in small woodwork, and is divided into many types, including flying cover, floor cover, balustrade cover and several leg covers.

The flying hood is used to form an archway between two pillar doors, but does not fall to the ground. The hood is often made of hollowed out wooden lattice or carved panels, using relief and translucent carving techniques to show the ancient, exquisite, quiet, elegant and clean artistic effect, its pattern is mostly geometric patterns or intertwined plants and animals, or mythological stories and so on (Figure 7–17).

The floor coverings are a type of carved wooden coverings for the interior eaves of ancient buildings. Any flower cover from the ground all the way to the beam (or square) can be called a floor cover (Figure 7–18).

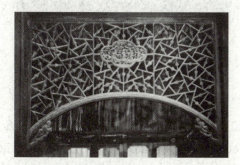

Figure 7–17 The flying hood

Figure 7–18 The floor coverings

A railing cover is a cover with a railing in the middle (Figure 7-19).

A few legs hood has small hanging columns on both sides, and the column head is not on the ground, especially like the legs of Zhuoji, so it is called a few legs hood (Figure 7-20).

Figure 7-19　Railing cover　　　　Figure 7-20　A few legs cover

2. Interior eave decoration—algae well

The algae well is the most beautiful part of the interior structure of the building, and can be described as a stunning sight to meet when you look up. The algae well is a component that covers the top inside the building, and is also known as algae well because of the patterns, carvings, and colorful paintings. The algae well has the craftsmanship of ingenuity and the prayer of wishing to know with the sky, which is the most beautiful sky in the eyes of Chinese people (Figure 7-21, Figure 7-22).

Figure 7-21　Algae well（1）　　　　Figure 7-22　Algae well（2）

3. Interior eave decoration—Wooden staircase

Wooden staircases in ancient buildings are used as vertical traffic elements between floors in buildings, mainly for traffic links between floors and when the height difference is large (Figure 7-23).

Figure 7-23 Wooden staircase

7.2 Interpretation of the spirit of traditional Chinese culture

The beauty of oriental rhythm can be glimpsed in the wood decoration of ancient buildings, where every carved beam and painted column and every secluded latticed window blossoms with the beauty of the craftsmanship of the ancients, the ingenious flower coverings and the most beautiful sky in the eyes of the Chinese, which makes people enchanted. The echoes of architectural notes among the jade buildings are the solidified spirit of the great craftsmen of China for 5,000 years.

7.3 Project innovation

Review the content of this section of the course, talk about which part of the wood decoration in ancient buildings impressed you, record your feelings to form a PPT report assignment.

Project 8   Color painting

8.1　Theoretical guidance

8.1.1　Grades of color painting

Color painting has a long history in China and is one of the most prominent features of ancient traditional architectural decoration. With its unique style and unique production technology and its magnificent decorative art effect, it has left a deep impression on people and has become the essence of traditional architectural art and entered the history books.

The color painting is also graded in ancient architecture, and there are three grades in total. The highest grade is HeXi color painting, followed by Spinzi color painting, and finally Su style color painting.

8.1.2　Characteristics of color painting

1. HeXi color painting

The main feature of the HeXi color painting is that the middle of the picture is made up of various dragon or phoenix patterns, and is complemented by floral patterns; the two sides of the picture are framed with 〈 〉 and drained and pasted with gold, which is very magnificent.

The most striking feature of the painting is the dripping and gilding, and the painting uses a large amount of gold, which is what we usually call gold and silk.

The main use of the HeXi paintings was for the important buildings of the outer court of the Forbidden City and the higher-ranking palaces where the emperor and empress lived in the inner court. Taihe Temple, Qianqing Palace, Yangxin Palace and other palaces use "gold dragon and seal painting" (Figure 8–1, Figure 8–2).

Figure 8-1　HeXi color painting（1）   Figure 8-2　HeXi color painting（2）

2. Spinzi painting

This painting is second in rank to the Hexi painting. The picture is painted with simplified form of scrolling petals, sometimes also can be painted with dragons and phoenixes, both sides are framed with 〈　〉, and can be pasted with or without gold powder. Generally used in secondary palaces or temples. The swirl color painting is mostly in lime green (Figure 8-3, Figure 8-4).

Figure 8-3　Spinzi painting（1）   Figure 8-4　Spinzi painting（2）

3. Sustyle color painting

Sustyle color painting originates from the traditional folk practice in the Suzhou and Hangzhou area of Jiangnan, hence the name, commonly known as "Suzhou piece". It is composed of two parts: pattern and painting, mainly used in classical garden architecture, giving people a lively, elegant, interesting and infinite reverie. Sustyle color painting has a wide variety of subjects, including flowers and birds, landscapes, figures, stories, etc (Figure 8-5, Figure 8-6).

Figure 8-5 Su-style color painting (1)   Figure 8-6 Su-style color painting (2)

8.1.3 The role of color painting

Color painting has a protective role, and the second is to play a decorative role. The original color painting is for the wooden structure moisture-proof, anti-corrosion, moth-proof, brush pigment on it to protect the wooden structure. After the Song Dynasty, color painting has become an indispensable decorative art for palaces, and is one of the most prominent features of ancient architectural decoration in China.

8.2 Interpretation of the spirit of traditional Chinese culture

As an inherent architectural culture of China, color painting shines in the flood of history. It conveys the culture and spirit of the long history of the Chinese nation, expresses different emotions and symbolic meanings, and reflects the accumulation of five thousand years of Chinese culture. The color painting has been with the ups and downs of history, but it has always been with the past, and it has always been new. In this long historical trajectory, the use of bright colors and auspicious patterns to express the auspiciousness and teach the world's moral meaning, it is a unique language to depict to us the wisdom, thought and aesthetics of our ancestors, expressing their expectations and visions for future generations, and is an important means and carrier for us to understand and learn about ancient culture.

8.3 Project innovation

Review the contents of this lesson, talk about your feelings through the restoration process of the color paintings and the color paintings, and record your feelings to form a PPT report assignment.

Project 9　Palace Architecture — Forbidden City

9.1　Theoretical guidance

9.1.1　The Concept of "Yangshi Lei Archives"

"Yangshi Lei Archives" is the name given to the Lei family that presided over the design of royal buildings during the Qing Dynasty for over 200 years.

Almost all of the important palace buildings and royal projects of the Qing Dynasty, such as palaces, tombs and imperial gardens, came from the Lei family. It is a treasure of Gan school architecture, which has flourished in the richness of the landscape and culture of the right side of the river, and has added a great deal of colour to Chinese ancient architecture.

9.1.2　The Origin of "Yangshi Lei Archives"

A southern craftsman, Lei Fada, came to Beijing to participate in the work of building palaces and was soon promoted to design work because of his high skills (Figure 9-1). From him, a total of eight generations until the end of the Qing Dynasty, the main imperial buildings such as palaces, imperial tombs, Yuanmingyuan, Summer Palace, etc. are the responsibility of the Lei family, this hereditary family of architects is known as "Yangshi Lei Archives".

During the past two hundred years, the Lei family has gone through eight generations of inheritance, from Lei Fada to Lei Xiancai (Figure 9-2).

Figure 9-1　Lei Fada

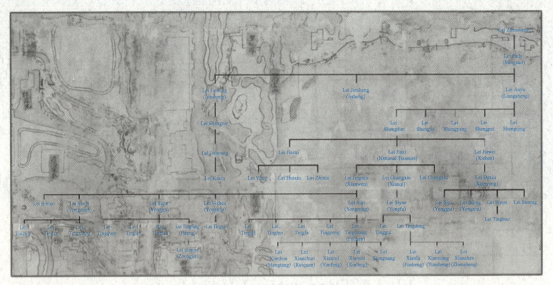

Figure 9-2　"Lei's Family" Chart

9.1.3　Classification of "Yangshi Lei Archives"

"Yangshi Lei Archives" is divided into drawing file and hot sample.

1. Drawing file

"Yangshi Lei Archives" drawings are drawn in ink, and the building dimensions and other construction instructions are indicated on the side of the building and site plans. According to the design procedure, it is divided into survey and site plan, architectural design sketch, in the presentation of the drawings, construction process; according to the projection method, it is divided into plan, section, elevation, drawing, perspective view (Figure 9-3—Figure 9-5).

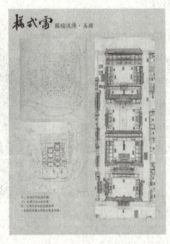

Figure 9-3　Drawing file of "Yangshi Lei Archives"

Figure 9-4　Perspective drawing file of "Yangshi Lei Archives"

Figure 9–5 Wood decoration drawings file of "Yangshi Lei Archives"

2. Ironing

Ironing sample is made of paper, Panicum and wood processing into the model drawing, because the final model ironed with a special small iron, so called ironing sample.

The ironing sample is divided into three categories: the first category is the full sub-sample, that is, the group of building ironing sample, a group of models (Figure 9–6, Figure 9–7); the second category is the sub-sample, that is, single building ironing sample, single model (Figure 9–8, Figure 9–9); the third category is the detailed sample, refers to the floor cover, screen and so on (Figure 9–10).

Figure 9–6 Full sub-sample of "Yangshi Lei Archives" (1)

Changchun Gong Hot Sample
Forbidden City 2009.8

Figure 9–7 Full sub-sample of "Yangshi Lei Archives" (2)

248 Appreciation of Chinese Traditional Architecture

Figure 9-8   Sub-sample of "Yangshi Lei Archives" (1)

Figure 9-9   Sub-sample of "Yangshi Lei Archives" (2)

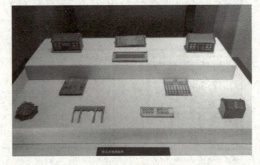

Figure 9-10   Detailed sample of "Yangshi Lei Archives"

When the roof of the hot sample is opened, the interior of the building can be seen, such as the structure of the beams and the style of color painting on the inner eaves. The label on the roof reads: "A building of Di'anmen, with a wide face of seven rooms, a width of 11 square meters and 4 feet 2 inches, and a depth of 3 square meters and 7 feet 6 inches from north to south. The face of the bright room is 2 square meters wide, the face of the second room is 1 square meter 7 feet 4 inches wide, and the face of the tip room is 1 square meter 5 feet wide. The height of the eaves column is one square meter and the diameter is one foot and eight inches. The middle column is two square meters and four feet and two inches high. Nine-purlins hutted roof, the bucket section single ang". In other parts of the hot stamping samples, there are also labels indicating the detailed dimensions and the names of the components. Through detailed observation of the hot stamping samples, it is possible to grasp the basic situation of the building from the whole to the details of Di'anmen.

9.2 Interpretation of the spirit of traditional Chinese culture

The term "architect" is the term of the day, and the term "craftsmanship" is also preached today. Although few craftsmen have gone down in history, the legendary stories of the fabulous

Lu Ban have been passed down from mouth to mouth.

These small models, created from paper and wood, still play their part as an important basis for the repair or restoration of some ancient buildings. To this day, the "Yangshi Lei Archives" family and their work continue to influence and move us.

9.3 Project innovation

Talk about your understanding of the spirit of artisanship in the context of this lesson, and record your thoughts and understanding in any form.

Project 10　Palace architecture — the three great halls of the outer court

10.1　Theoretical guidance

10.1.1　The name of the "three great halls of the outer court"

　　Beijing Imperial Palace buildings are divided into two parts: the outer court and the inner court, the center of the outer court is the Taihe Hall, Zhonghe Hall, Baohe Hall, collectively known as the Three Great Halls, is where the country held a large ceremony. The center of the inner court is the Qianqing Palace, the Jiaotai Palace and the Kunning Palace, collectively referred to as the Hou San Gong. It is the main palace where the emperor and the empress live.

　　The outer court is where the emperor deals with political affairs, of which Taihe Hall is the tallest and most splendid, where grand ceremonies are held when the emperor is enthroned, married, enthroned, ordained, and marched.

10.1.2　Taihe Hall

1. Overview of Taihe Hall

　　Taihe Hall is the largest and highest-ranking building in the Forbidden City is one of the three great halls of the East, the largest surviving wooden structure of the Great Hall in China. Located in a prominent position on the north-south axis of the Forbidden City in Beijing, it has a wide face of eleven rooms, a depth of five rooms, 64.24 meters long and 37 meters wide, with a floor area of 2,377.00 square meters, 26.92 meters high, together with a pedestal height of 35.05 meters, with 10 animals placed in the corners of the eaves, and a wide platform called the Dan Dynasty, commonly known as the moon platform, the largest hall in the Forbidden City.

　　Everything about the structure of the Hall of Harmony conveys a message — the emperor is supreme, and all harmony is centered on him (Figure 10–1).

Figure 10-1 Taihe Hall

2. Function of Taihe Hall

There are many people think that the Taihe Hall is usually used for the court, in fact, it is not. Taihe Hall is actually used to hold major ceremonies, the actual number of times used, the Ming and Qing emperors on the court mainly in the Taihe Gate, Qianqing Gate (for the imperial door to listen to politics), Qianqing Palace (there are major or important things when the emperor summoned ministers location), and Yangxin Hall (late Qing dynasty draped the government), and not the usual Taihe Hall. The 24 emperors of the Ming and Qing dynasties were held in the Hall of great ceremonies, such as the Emperor's enthronement to the throne, the Emperor's wedding, the empress, the order will go to war, in addition to the annual Wanshou Festival, New Year's Day, the winter solstice three major festivals, the Emperor here to accept the congratulations of civil and military officials, and to give feasts to the princes and ministers.

3. Interior decoration of Taihe Hall

Taihe Temple all use gold bricks paving, but the floor of the 4,718 gold bricks are not really made of gold, but a special brick from Suzhou, because when struck can make a metallic sound so called gold bricks. Therefore, Taihe Temple is also known as the "Golden Temple" (Figure 10-2).

Taihe Hall interior and exterior decoration is extremely luxurious, the beams and pillars are painted and seal color painting, the upper part of the doors and windows are diamond lattice pattern, the lower part is carved in relief with cloud and dragon patterns, the joints are engraved with gilt bronze leaves with dragon patterns (Figure 10-3).

Figure 10-2 Golden tile pavement inside Taihe Hall

Figure 10-3 The interior of Taihe Hall is luxuriously decorated

10.1.3 Zhonghe hall

Zhonghe Hall is the emperor went to Taihe Hall before the ceremony to take a short break and practice the rituals of the place, the emperor before going to Taihe Hall stop here to accept the cabinet ministers and officials of the Ministry of Rites salute, and then into Taihe Hall for the ceremony (Figure 10-4).

In addition, the emperor sacrificed to heaven and earth and the temple before, but also here to review the written sacrifice of the "wish plate", in the South China Sea before the performance of plowing, but also here to review the plowing tools, Zhonghe hall is located between the Taihe hall and the Baohe hall, so it has a middle meaning.

The word "Zhonghe" means that everything should be done impartially and appropriately in order to use the relationship between all parties to get harmony, which is meant to promote the "middle way".

10.1.4 Baohe hall

Baohe Hall is the annual New Year's Eve the emperor gave a feast on the place of the feudal princes, the Ming Dynasty ceremony before the emperor often dress here, the empress, the crown prince, the emperor was congratulated in this hall, Baohe hall is also the imperial examinations held in the temple test place (Figure 10-5).

Figure 10-4　Zhonghe hall

Figure 10-5　Baohe hall

10.1.5 The "Thousand Dragons Spitting Water" in the Three Great Halls

There is also a hidden wonder in the Three Great Halls, namely the "Thousand Dragons Spitting Water". The Thousand Dragons are more than a thousand stone faucets sticking out from underneath the columns, and the rainwater is discharged from the mouths of the dragons whenever it rains, making the practical function of diverting rainwater and the ornamental function of architectural art organically combined (Figure 10-6).

The spectacular sight of a thousand dragons spitting out water comes from the three-story pedestal on which the three halls of Taihe, Zhonghe and Baohe stand.

Figure 10-6 "Thousand Dragons Spitting Water" in the Three Great Halls

10.2 Interpretation of the spirit of traditional Chinese culture

The three halls are magnificent and can be considered the pinnacle of art in Chinese architectural history, with each unique design incorporating the soul of Chinese artisans and the spirit of Luban. The carved beams and painted columns are permeated with the marks of history, showing the magnificence and vigor of the nation. It is the duty of each and every one of us to promote the spirit of the nation by absorbing the spirit of heaven and earth and sucking the essence of the sun and the moon.

10.3 Project innovation

Write your own experience with the content of this section of the course and incorporate where the spirit of the Chinese nation.

Project 11　Palace architecture — ten animal ornaments on house-ridges

11.1　Theoretical guidance

11.1.1　The overview of "ten animal ornaments on house-ridges"

The "ten animal ornaments on house-ridges" are sculptures placed on the ridges of houses and palaces in ancient Chinese traditional architecture. There are up to ten running beasts on ancient Chinese buildings, which are distributed on the hanging ridges at both ends of the houses. The "Ten Beasts of the Roof" were placed on top of the highest-ranking Taihe Hall in the Forbidden City.

11.1.2　The name of "ten animal ornaments on house-ridges"

"Ten animal ornaments on house-ridges" are from top to bottom, the Immortal, the Dragon, the Phoenix, the Lion, the Sea Horse, the Heavenly Horse, the Apocrypha, the Mythical Creature, the Haetu, the Bull and the Xing Shi. The ten beasts on the roof were mentioned earlier, but why are there eleven here? The immortal here is the leader of the ten divine beasts that follow. In ancient times, divine beasts were placed on the ridge of the roof according to the status and rank of the owner, which is why the Hall of Taihe, representing the highest level of imperial power, has these ten divine beasts (Figure 11-1).

11.1.3　The meaning of "ten animal ornaments on house-ridges"

1. The Immortal

The first is the Immortal, also known as the Immortal Riding the Phoenix. His function is to hold the tiles beneath the hanging ridge, but he also marks the development of Chinese architectural aesthetics, and his symbolism is that he can fly on the phoenix and turn good fortune into good fortune. At the same time he is a warning to the world that if they climb any further they will fall to pieces (Figure 11-2).

2. The Dragon

The dragon, also known as the owl, signifying supreme imperial power. In ancient times, the dragon was an auspicious object to avoid fire. To this day, one can still see the "dragon culture" displayed in new antique buildings (Figure 11-3).

Figure 11-1   The name of ten animal ornaments on house-ridges

Figure 11-2   The immortal

Figure 11-3   The dragon

3. The Phoenix

The phoenix is used in traditional culture as a metaphor for a person of holy virtue, the so-called holy virtue referring to the "five constants", namely benevolence, righteousness, propriety, wisdom and trust, as proposed by Dong Zhongshu in the Western Han Dynasty. The phoenix is also a symbol of the emperor's noble status (Figure 11-4).

It is a symbol of beauty and good fortune, and is a symbol of Chinese culture and spirituality. Like the dragon, the phoenix is not an animal that exists in reality, but is a divine object that is a fusion of many animals that live in reality. China has always had the beautiful symbolism of the dragon and phoenix.

4. The lion

The lion depicted here is a Chinese translation of the lion, whose symbolic meaning is one of courage and majesty. The lion is the king of the beasts, and when it roars, the beasts will cower in fear, representing bravery and majesty. The lion's fierce and arrogant appearance has a deterrent effect on some demons and monsters. The lion can help to protect the peace of the home. The stone lion is like a guardian deity, protecting the home from evil spirits (Figure 11-5).

Figure 11-4　The phoenix

Figure 11-5　The lion

5. The Sea Horse

The sea horse, also known as the Dragon's Son, symbolises loyalty, courage and good fortune, and its wisdom and virtue reach all directions through the sky and into the sea. The seahorse enters the sea and the abyss to turn good fortune into good fortune, and is a faithful and brave beast in ancient Chinese mythology. The seahorse likes to walk through the waves, and his body is covered with a pattern of vibrating waves and rocks, which signify the sea and river cliffs, and the pattern on the seahorse is also representative of ancient auspicious motifs (Figure 11-6).

6. The Heavenly Horse

The idiom we know as Tianma Xingkong refers to the heavenly horses here. A heavenly horse can travel a thousand miles a day, as the saying goes. Coming alone, he can be proud of the herd and open up frontiers, and its symbolism is that he can turn evil into good fortune (Figure 11-7).

7. The Apocrypha

In the sea, the Apocrypha is the leader of the aquatic fishes and is a special beast in the sea. According to legend, both the Apocrypha and the mythical animal are gods who can make clouds and rain, put out fires and prevent disasters. It is said to be able to emit columns of water, implying that it can make rain and wind, and its symbolic meaning is that it can prevent fires and disasters (Figure 11-8).

8. The Mythical Creature

The mythical animal is good at making rain and wind, although fierce in nature, but good in heart, representing good luck. It resembles a lion, and likes to sit quietly but not to move, and it likes fireworks (Figure 11-9).

Figure 11-6　the sea horse

Figure 11-7　the heavenly horse

Figure 11-8　The Apocrypha

Figure 11-9　The mythical animal

9. The Haetu (Xie zhi)

Xiezhi has a horn on its head and is powerful and good at judging evil and righteousness, so it is regarded as a symbol of law. It is said in the book of Exotic Objects that "there is a beast in the northeast desert named Xiezhi". A horn, the nature of loyalty, see people fighting is not touching the straight, hear people's argument is not correct. It can distinguish the right from the wrong, and it is also known as the "sacred sheep", which is a symbol of courage and justice (Figure 11-10).

10. The Bull

The bull has the same role as the dragon, hornless and fish, and they are the same water-suppressing beast. It is often placed in places where flooding occurs, so it has the meaning of calming evil and protecting homes (Figure 11-11).

The fighting bull, with its horned dragon body and dragon claws and four fingers, is often found on the third-rank official uniform rewarded by the Ming Dynasty, called the fighting bull uniform. The form is a smaller grade of prisoner bull. It is a sacred animal between the Dou and the Ox.

11. Xing Shi

It is called Xing Shi because it ranks tenth among the ten divine beasts. It holds a vajra pestle in its hand, which resembles the image of the legendary thunder god, so it is often placed on the roof to protect against lightning (Figure 11–12).

It also has another symbolic meaning that it can help you keep your secrets, just like our sidekick's best friend or brother, so it can help people soothe their emotions.

Figure 11–10 The Haetu　　Figure 11–11 The Bull　　Figure 11–12 Xing Shi

11.2　Interpretation of the spirit of traditional Chinese culture

Each of the sacred animals is the crystallization of the wisdom and creativity of the ancients, and we cannot help but marvel at these supreme artistic forms, guarding the artistic treasures that unite the long historical and cultural accumulation and the spiritual wisdom of the nation. Through the understanding of the ten beasts of the roof, we appreciate the profound culture of the Chinese nation, so we have to take the long history of the nation and the fine culture and traditional moral cultivation Therefore, we should carry forward the long history of the nation and the fine cultural and traditional moral cultivation.

11.3　Project innovation

Review the knowledge points and historical elements related to the ten beasts of the roof in this lesson, such as the auspicious patterns on the sea horse , and complete the PPT summary report assignment after the lesson.

Project 12 Palace architecture — the aesthetics and symbols of color in the architecture of the Forbidden City

12.1 Theoretical guidance

12.1.1 First impression of color in the Forbidden City

Many people's impression of the color of the Forbidden City buildings are red walls, yellow tiles, vermilion doors with gold nails, and occasional lime green colored paintings decorating under the roof (Figure 12-1—Figure 12-4).

Figure 12-1 "Red Wall" in the Forbidden City

Figure 12-2 "Yellow tiles" in the Forbidden City

Figure 12-3 "Golden nails" in the Forbidden City

Figure 12-4 Painted decoration in the Forbidden City

12.1.2 Aesthetic interpretation and symbolism of colour in the Forbidden City

We interpret these architectural colours on five levels: tiles, eaves, columns and walls, pedestals and balustrades, and ground.

1. Tiles

The main palaces on the central axis of the Forbidden City are covered with yellow tiles, while the three southern palaces to the east are covered with green tiles. The twelve palaces to the west are covered with golden yellow tiles, while the Wu Men platform to the south is covered with red and the Wen Yuan Ge to the north is covered with black tiles (Figure 12-5).

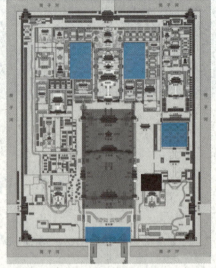

Figure 12-5 Diagram of the colour distribution of the "Palace tiles"

(1) Yellow glazed tile. In *Zhou Yi-Kun*, it is said that "the sky is mysterious and the earth is yellow", which means that the universe is profound and the earth is nurtured. In Chinese traditional culture, the earth is yellow, so yellow is used to represent "earth". Since land is a symbol of the state, yellow also represents the power of the emperor.

In the Book of *Shang Shu*, there are five elements of yin and yang, namely gold, wood, water, fire and earth. Among them, yellow corresponds to "earth", and it also represents lightness and liveliness.

There are 103 tiles on the two layers of the roof in the Hall of Harmony, each tile is a work-like tile, one has seven or eight pounds of weight. Such a heavy roof because of the yellow roof does not seem so heavy, but more light and lively. The yellow glazed tile symbolizing the royal power is also called "golden coat" (Figure 12-6, Figure 12-7).

Figure 12-6 Yellow glazed tile

Figure 12-7 The building covered with yellow glazed tile

The roofs of the common people are not allowed to use yellow tiles, their roofs are usually black, black tiles are also called cloth tiles, so the people in ancient times are also called "cloth".

(2) Green glazed tile.The green tile is located in the east, and the east represents the place where the sun rises. In the five elements of yin and yang, the east is wood, green, the main growth. So the residence of the imperial son in the east, covered with green tiles means that the imperial son can thrive.

The eastern part of the Forbidden City was mainly where the imperial sons studied and lived, so the "three southern houses" building was distinctive (Figure 12-8, Figure 12-9).

Figure 12-8   Green glazed tile

Figure 12-9   The three southern houses covered with "green tiles"

(3) Golden Glazed Tile.The west is where the sun sets, while the golden color signifies golden autumn. In golden autumn, all things will be fruitful. In the five elements of yin and yang, west is gold, so the west area is divided into functional areas for the empress and the concubines to live in, for the empress to spend her old age here and have a successful life; for the concubines, it means that they can give birth to more children for the emperor and bear the fruit of life, which is conducive to the prosperity of the royal offspring and many sons and daughters (Figure 12-10, Figure 12-11).

Figure 12-10   The Harem covered with "golden tiles"

Figure 12-11   "Golden glazed tiles" in the Forbidden City

(4) Red Noon Gate. The south is covered in red, the south mainly refers to the Noon Gate, the roof of which is covered in yellow tiles, but its bearing platform is indeed red. In the five elements of yin and yang, the south is red for fire. The high surface of the platform is decorated in red, which is majestic and solemn and sets off the height and majesty of the Gate (Figure 12-12).

(5) Black glazed tile. The building covered in black tiles is called Wen Yuan Ge, and it functioned as a book collection and the emperor's private library. As the saying goes, water and fire are incompatible, and water is used to counteract fire.

Black is the colour of water in the five elements of yin and yang, so the tiles on the roof of Wenyuange were covered in black to symbolise the need to keep fire away (Figure 12-13).

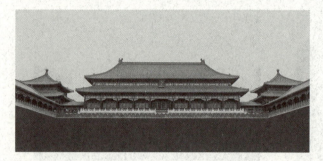

Figure 2-12 Noon Gate

Figure 12-13 Wen Yuan Ge covered with "black tiles"

2. Eaves

The eaves mainly refer to the frontal square and arch, and their color is green. Green makes people think of nature's green trees and green grass, giving people a sense of spring, a sense of vitality; blue makes people think of the sky and the sea, feeling the vast expanse of heaven and earth, the tranquility of nature.

As the eaves are raised outward, shadows will appear in the lower part of the beams and arches. The cold color, which looks airy, light and distant in the shadow, makes the heavy roof give people a sense of relaxation, and enhances the sense of height and space of the building, with the feeling of expanding space (Figure 12-14).

3. Columns and walls

The color of the pillar frames and walls in the Forbidden City is red, which is also one of the main colors of the ancient architecture of the Forbidden City. Red gives people hope and satisfaction, since ancient times, folk will also be red as the color of celebration. The Ming and Qing dynasties stipulated that all zhengzhang sent exclusively to the emperor must be red, indicating the emperor's love for the color red.

Functionally speaking, the walls play a maintenance role for the building, while the columns are important components to support the roof of the building. It can be seen that Both components can play a defensive and protective role for the building. Its color is red, which is conducive to reflecting the masculine spirit and guarding Royal architecture (Figure 12–15).

Figure 12–14 The color of the frontal square and arch in the Forbidden City

Figure 12–15 Red pillars and walls in the Forbidden City

4. Pedestals and balustrades

The pedestal and balustrade are white, which is a symbol of elegance, purity and nobility. As the Forbidden City ancient building pedestal railings and pillars have exquisite dragon and phoenix carvings, so the use of white Chinese white jade material is conducive to highlight the nobility of the building itself.

At the same time, the white pedestal and yellow roof and red columns form a sharp contrast, which can show the magnificence and elegance of the building in one (Figure 12–16).

Figure 12–16 Table base and railing

5. Ground

The color of the ground of the Forbidden City ancient buildings outdoor ground is gray, the interior of the room is mostly gray and black (gray is the original color of the brick, black for the ground when the plastering is almost complete in the surface of the black alum water splashed) (Figure 12–17, Figure 12–18).

Figure 12–17 Outdoor floor

Figure 12–18 Indoor floor

From the perspective of location and function, the palace building floor should not use bright colors, thus using gray-black with low-key characteristics, this color is located in the middle of various shades, it is silent and receding with various colors and fused with various shades, forming a good complementary color effect. At the same time, compared with the white of the table base railings, gray-black and white form a sharp color difference It gives life to white, which is also an intermediate color.

The relaxed black and the circumscribed white make the color of the Forbidden City tend to be more perfect.

12.2 Interpretation of the spirit of traditional Chinese culture

The Forbidden City ancient architecture of the clever use of color, not only reflects the culture and art of ancient architecture itself, but also reflects the wisdom of our ancient artisans thinking.

It can be said that the symbol of architecture to a certain extent carries the culture of a country or nation, in the five thousand years of Chinese history and civilization, China's traditional outstanding cultural accumulation has not been lacking in the figure of many many buildings this important role, and color as an important component of the building, both to cause visual feedback of all kinds, but also has a shocking rendering power.

The yellow glazed tile roof of the Forbidden City in Beijing is glorious, the vermilion columns and doors and windows, the green and slightly gold architectural paintings in the shadow part under the eaves, against the white pedestal, make the outline of each part of the building more distinctive, with a unique color. The different colors of each ministry are the embodiment of the feudal social hierarchy and represent the internalization of feudal rituals in architecture.

12.3 Project innovation

Review the knowledge points in this section of the course and draw a mind map based on what was covered.

Project 13   Small House Gate — the quadrangle

13.1   Theoretical guidance

13.1.1   The classifications of quadrangle

A courtyard is a traditional Chinese courtyard-style building with houses on all sides. The courtyard is enclosed in the middle from all sides. The courtyard is classified according to the number of courtyards, which can be divided into one courtyard, two courtyards, three courtyards and four or five courtyards.

First of all, if the courtyard is in the shape of the word "口", it is called one courtyard, and the whole courtyard is in the shape of the word "口"; the one in the shape of the word "日" is called two courtyards; the one in the shape of the word "目" is called three courtyards; the four courtyards The courtyard is divided into four sections; the existing five-entry courtyard is very rare. The complete courtyard is a three-entry courtyard, and it is also the courtyard form that we need to focus on.

From the front to the back, there is the gate of the house, the inverted seat room next to the gate, followed by the shadow wall, the pendant door behind the shadow wall, then the corridor attached to the pendant door, the two sides of the compartment, then the main house, the ear room on both sides of the main house and the innermost back room (Figure 13-1).

13.1.2   The system of house gate

The gate of the house refers to the gate of the courtyard facing the street, which is an apparent characteristic of the status of the master in the old society, and the gate of this is generally divided into the gate of Wangfu, the gate of Guangliang, the gate of Jinzhu, the gate of Manzi and the gate of Ruyi.

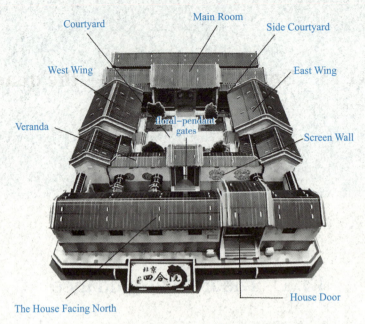

Figure 13-1 "Three courtyards" layout of courtyards

1. Wangfu Gate

Wangfu gate is the highest level of the courtyard door, is the exclusive use of the royal family door. The most common two specifications are five rooms with three doors and three rooms with one door (Figure 13-2).

2. Guangliang Gate

The Guangliang Gate is the most representative gate of Beijing courtyard, and it is a gate that can only be used by officials of a certain rank. The pedestal of Guangliang Gate is high, and the main body of the gate structure is the front and rear eaves pillars and the middle pillar (Figure 13-3).

Figure 13-2 Wangfu Gate

Figure 13-3 Guangliang Gate

3. Jinzhu Gate

Jinzhu Gate is a gate that can only be used by official and eunuch families of a certain rank. The gate is set on the gold pillar between the middle pillar and the front gable pillar, and compared to the wide space in front of the gate of Guangliang, the gate of Jinzhu has less space in front of the gate (Figure 13-4).

4. Manzi Gate

It is said that the door is placed directly on top of the front gable in order to avoid thieves from staying in front of the door. The name "Manzi Gate" was derived from the fact that the ancient Chinese word for "Manzi" means "rough" and "savage" (Figure 13-5).

5. Ruyi Gate

Ruyi Gate is the most commonly used gate in Beijing Courtyard, which is built between the front eaves pillars and has two door pins at the entrance of Ruyi Gate, mostly engraved with the word "Ruyi", which is also the origin of the name of Ruyi Gate (Figure 13-6).

Figure 13-4 Jinzhu Gate   Figure 13-5 Manzi Gate   Figure 13-6 Ruyi Gate

13.1.3 Shadow wall system

The shadow wall, also known as the shade wall, is a wall used in traditional Chinese architecture to block the view and increase the momentum of the house, and ancient people attached great importance to protecting privacy.

1. One-character shadow wall

Some of them are too long and look too monotonous, so they are divided horizontally into three sections, with a long middle and two short ends, forming a main and two subordinates (Figure 13-7).

2. eight-character shadow wall

The eight-character shadow wall on the basis of a word shadow wall, the two from the inward fold, forming a eight-character enclosed shape (Figure 13-8).

3. Block mountain shadow wall

Block mountain shadow wall is built directly on the wall of the compartment small wall cap and make shadow wall shape, so that the shadow wall and the wall of the mountain as one (Figure 13-9).

4. Skimming mountain shadow wall

Skimming mountain shadow wall is the "anti-eight-character shadow wall", when doing this kind of shadow wall, the door should be backward 2 to 4 meters, in front of the door to form a small space, can be used as a buffer to enter and exit the door, which is also the most obvious difference with the eight-character shadow wall (Figure 13–10).

Figure 13–7   One–character shadow wall Figure

Figure 13–8   Eight–character shadow wall

Figure 13–9   Block mountain shadow wall

Figure 13–10   Skimming mountain shadow wall

13.1.4   The name and function of each component of the quadrangle

1. House Gate

The gate of the house is the gate of the courtyard facing the street, which is an apparent characteristic of the status of the master in the old society.

2. Inverted seat room

The inverted seat room is connected with the gate of the house. The inverted seat room is adjacent to the gate of the courtyard, and it is the house facing north and south opposite to the main room, also called the south room, which is usually the outer living room, the account room, the concierge, and the guest room, so it is mostly used for guests or servants to live (Figure 13–11).

3. Shadow wall

The shadow wall is also known as the wall of light is used in traditional Chinese architecture to block the view and increase the momentum of the house.

4. The pendant door

After passing through the shadow wall, you will see the pendant door, which is the boundary line and the only channel between the inner and outer houses, and after entering this door is the interior of the courtyard, outsiders are generally not allowed to enter and exit casually. This is one of the characteristics of the Weeping Flower Gate, the eaves and pillars of which are not on the ground, and the whole building occupies the sky but not the land (Figure 13–12).

Figure 13–11   Inverted seat room

Figure 13–12   The pendant door

5. Copying hand corridor

The copied hand corridor is also called corridor, copied hand corridor is a common form of corridor in the courtyard, and the pendant door is connected, around the pendant door, the compartment and the main room, it is designed to facilitate people to walk in the rain and snow (Figure 13–13).

6. Compartment room

The compartment is the house on the east and west sides of the courtyard, divided into east and west compartments, usually the eldest son lives in the east compartment, the second son lives in the west compartment (Figure 13–14).

7. Main house

The main room is the core building of the courtyard, the residence of the head of the family, generally three rooms, the bright room of the main room, that is, the middle of a room called the hall, also known as the hall; three open rooms of the main room on both sides of the hall generally for the bedroom and study, the main room is characterized by the sun can shine into the house in winter, warm in winter and cool in summer (Figure 13–15).

Figure 13–13   Corridor

Figure 13–14   Compartment room

Figure 13–15   Main house

8. Ear room

There are one or two rooms on each side of the main house with small depth and height, just

like two ears hanging on both sides of the main house, so it is called ear room. The function of the ear room is that each can be used as a warehouse and kitchen (Figure 13-16).

9. Back room

The back cover room is behind the north room, there is one layer, there are also two layers, are sitting in the north facing south, and the back wall of the north room and between the formation of a courtyard, generally known as the backyard, the backyard is for the house owner's family or the elderly to live (Figure 13-17).

Figure 13-16   Ear room

Figure 13-17   Back room

13.2   Interpretation of the spirit of traditional Chinese culture

The quadrangle is generally a single-family house, but there are also cases of multiple families sharing a single quadrangle, mostly poor families, which we call "dachaoyuan". Although the courtyard is a residential building, it has a profound cultural connotation and is a carrier of traditional Chinese culture. The decoration, carving and painting of the courtyard reflect the folk customs and traditional culture, and express people's pursuit of happiness, beauty, affluence and good fortune under certain historical conditions.

13.3   Project innovation

Draw a "three courtyard house" and label each section with its name in English and Chinese.

Project 14　Great House Gate — Qiao Family Compound

14.1　Theoretical guidance

14.1.1　The basic architectural system of the Qiao Family Compound

Qiao family compound, also known Zaizhongtang, is located in Qiaojiabao Village, Qixian County, Shanxi Province. First built in 1756, the entire compound is in the shape of a double "喜", divided into six courtyards, set of 20 small courtyards, a total of 313 houses, with a construction area of 4,175 square meters. Qiao family compound is surrounded by a fully enclosed brick wall of more than 10 meters high on three sides, and is an ancient house with traditional northern residential architectural style.

14.1.2　The six courtyard layout and architectural features of the Qiao Family compound

1. The First Courtyard

The first courtyard, also known as the Old Courtyard, was built by Qiao Zhiyong's father, Qiao Quanmei.

After entering the gate from the first courtyard, there is a wall called the Fude Ancestral Hall, which proves that "the door welcomes a hundred lives, and the courtyard receives good fortune", and its purpose is for decoration and to calm down the house to avoid suspicion. The first courtyard has a plaque on the front door, on which is engraved the words "No disrespect", from the first sentence of *Qu Li*, the first part of Zhu Zi's Book of *Rites*, which we can understand that there are gods and spirits between heaven and earth, and an attitude of disrespect is not desirable (Figure 14–1).

2. The Second Courtyard

The second courtyard exhibits the history of the Qiao family. The courtyard here is built in the form of a "double happiness", and the plaque in front of the door is engraved with "Huifang", which means that virtuous people gathered together. The couplet below the plaque is engraved with the words "Be magnanimous and frank, the family will always be prosperous; be gentle and peaceful, the family will always be rich; be gentle and peaceful, the family will always be glorious and the descendants will be prosperous". This fully reflects the Qiao family's strict family rules and the spirit of Jin business (Figure 14–2).

Figure 14–1　The First Courtyard of the Qiao family

Figure 14–2　The Second Courtyard of the Qiao family

3. The third courtyard

The third courtyard is also called the three treasure courtyards, which conceals three town treasures.

（1）Ten Thousand People's Ball. The Ten Thousand People's Ball is a mercury glass ball hanging from the ceiling, which is said to have been purchased by the Qiao family from the United States. This glass ball will shrink the image of the people in the house according to a certain proportion and reflected in the ball, is when the Qiao family meeting guests to talk business hung on the room to monitor the people in the house, also known as the world's first surveillance probe, for the time extremely rare things (Figure 14–3).

（2）Rhinoceros Moonscope. Rhinoceros looking at the moon mirror is a diameter of about one meter mirror, mirror frame frame for the hardest wood of rare trees made of ironwood, two meters high and weighing one ton, the whole treasure consists of three parts: the top is the mirror on behalf of the round moon; the middle part is the auspicious clouds.

The base is the rhinoceros, where "rhinoceros" is the harmonic sound of "happy cattle", is a great luck, happy from the sky, while the rhinoceros looking at the moon mirror is also classified as a national cultural relics (Figure 14–4).

（3）Nine Dragons Lantern. Nine Dragons Lantern is made of mahogany in the Qing Dynasty. At that time the Empress Dowager Cixi fled to Xi'an when the Eight-Power Allied Forces invaded China, the Qiao family donated 300,000 taels of silver to meet the Empress Dowager's urgent needs, after which the Empress Dowager Cixi rewarded these two Nine Dragons lamps for their loyalty, and no third one has been found in the country (Figure 14–5).

Figure 14–3 Ten Thousand People's Ball

Figure 14–4 Rhinoceros Moonscope

Figure 14–5 Nine Dragons Lantern

4. The fourth courtyard

The fourth courtyard is a commercial and secular courtyard, and a new courtyard. It was built by Qiao Yingxia, the grandson of Qiao Zhiyong. Because Qiao Yingxia had the experience of studying abroad in his early years, this courtyard reflects the advanced technology brought by his study abroad in every aspect of the architectural style. The most distinctive feature is the carving below the eaves of the east wing, which shows a train whistling on a bridge, drawn by Qiao Yingxia from his memory, as there were no trains in China back then (Figure 14–6).

The brick carvings above the lintel are engraved with a Chinese qin, chess, calligraphy and painting, meaning that the Qiao family wanted their children and grandchildren to be versatile and proficient in everything (Figure 14–7).

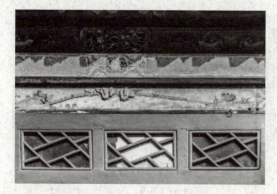
Figure 14–6 Carving below the eaves of the fourth courtyard

Figure 14–7 Brick carving above the lintel of the fourth courtyard

5. The fifth courtyard

The Fifth Courtyard, also called Zaizhongtang, was built by Qiao Zhiyong. First of all, we see a plaque above the lintel, engraved with the words "Zaizhongtang". It means "impartiality, impartiality, impartiality, impartiality", reflecting the owner's middle-of-the-road thinking; the lintel is also carved with three stars of happiness, longevity and prosperity, and there are carvings of eight stallions on both sides, which means that the owner hopes his descendants will make a difference.

The couplet below is engraved with the words "The only way to pass on the family line is to store thick, and there is no strange way to deal with the world but to be frank and sincere", which means that the only true way to pass on the family line is to accumulate thick virtue, and there is no strange way to deal with the world and establish oneself, as long as one follows the straightforward and sincere nature. Here it is fully explained that the Qiao family emphasizes generosity, speaks sincerely, does not deceive or cheat, does not conceal or deceive, and takes advantage of righteousness. Integrity gathers popularity and virtue remits wealth. The simple couplet illustrates the owner's way of passing on his family and dealing with the world (Figure 14-8).

6. The sixth courtyard

The sixth courtyard is where the owner of the Qiao family, Qiao Jian, originally intended to build a "study courtyard" and had already bought back a full set of rockery at that time. However, with the outbreak of the "July 7th Incident" in 1937, only a part of the garden was left here, and it was changed to the "Garden House", leaving the garden unfinished (Figure 14-9).

The study room was used to train the children of the Qiao family, hoping that they would study hard, grow in knowledge, and become outstanding talents, so the study room was adjacent to the ancestral hall to inspire the future generations of the Qiao family to progress in their studies.

Figure 14-8　Fifth courtyard
(Zaizhongtang)

Figure 14-9　Sixth courtyard
(in the garden courtyard)

14.2 Interpretation of the spirit of traditional Chinese culture

The Qiao family compound fully embodies the spirit of Chinese Jin merchant culture, where the confrontation between traditional culture and merchant culture is not seen. The Qiao family under the governance of Qiao Zhiyong is based on the principle that those who do business first establish themselves. The Qiao family under the governance of Qiao Zhiyong is based on the standard that only those who are loyal and righteous are businessmen, which further complements the traditional culture's theory of cultivating one's moral character, cultivating one's family and even governing the country, and combines self-cultivation, self-improvement and family spirit, pioneering and enterprising. In this way, many beautiful characters came — thrift, honesty, tolerance, self-restraint, and these also accompanied the family all the way to glory, even though the history of the development of twists and turns, natural and man-made disasters visited incessantly, and the family, always maintaining a high morale, in this beautiful and rich land, gathering and dispersing quicksand, the wind rose. The wind rises and falls.

14.3 Project innovation

Talk about your feelings, including your understanding of the spirit of Jin business culture. You can record your feelings in a short video, or form a 500-word written record.

Project 15　Huizhou Folk Houses

15.1　Theoretical guidance

15.1.1　The basic architectural form of Huizhou houses

Huizhou houses, also called Huizhou-style houses. It is an important school of Chinese traditional residential architecture, because Huizhou area is known as a natural park, so its architectural style and natural scenery are in harmony, which makes the practicality and artistry perfectly unified.

The basic architectural from of Huizhou Folk house are divided into "凹" layout, "回" layout, "H-shaped" layout and "日" layout.

1. "凹" layout

The "concave", triple courtyard type, also called three-room type, is the most economical one among Huizhou dwellings. It is mostly a two-story house with a patio in the center, a living room in the bright room downstairs, and an ancestral hall dedicated to the ancestors' gods in the bright room upstairs. The two side rooms can be used as bedrooms (Figure 15-1).

2. "回" layout

The "back shape", also known as "口" shape, quadrangle style, three rooms and two, the foyer upstairs bright room for the main room, back into the hall upstairs bright room as a general ancestral hall, both sides of the room is the bedroom (Figure 15-2).

3. "H-shaped" layout

"H" shaped, with two back-to-back triplexes in the front and back. There are three rooms and two halls, with a patio at the front and back. This kind of house has no foyer, and the main hall is often divided into two halls, and the two halls share a ridge, commonly known as "one ridge over two halls" (Figure 15-3).

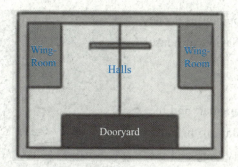

Figure 15-1 "凹" layout

Figure 15-2 "回" layout

4. "日" layout

The "日" layout has three rooms and three entrances. There are two patios, and the two sides of each room are connected by a corridor. The veranda refers to the veranda house, and the ancestral hall is usually located in the last hall (Figure 15-4).

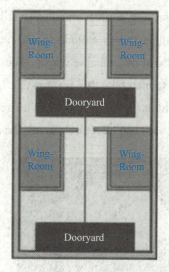

Figure 15-3 "H-shaped" layout

Figure 15-4 "日" shaped layout

15.1.2 The difference between Huizhou houses and courtyard layout

The difference between the southern courtyard and the northern courtyard in terms of architectural style, take the "three-in-one" style as an example, the northern courtyard is not a house-over-house layout, the biggest difference between the two is the patio. There is no patio in the northern courtyard, but we can see the existence of patio in each of the previous Huizhou residential layout (Figure 15-4).

15.1.3 The characteristics of Huizhou residential architecture

1. House over house

The whole architectural system of Huizhou houses is one house over another, forming the architectural layout of house over house (Figure 15-5).

2. Patio

The shape of the patio is generally rectangular, and the role of the patio can collect light and ventilation, because the walls around the Huizhou houses are very high and the windows are narrow, so people will use the patio to collect light and ventilation, and then it also has the function of drainage, so it has the auspicious meaning of "four waters return to the hall" (Figure 15-6).

3. Horse head wall

Horse head wall, also known as wind and fire wall, firewall, fire sealing wall. Horse-head wall refers to the top part of the wall, which is called "horse-head wall" because

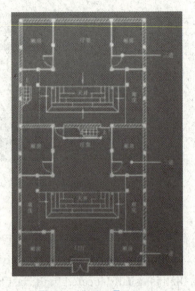

Figure 15-5 Features of the "house over house" architectural layout

of its shape resembling a horse's head. The main reason why the walls of traditional residential buildings in Jiangnan take this form is that in villages where people live together, the density of residential buildings is large, and when a fire occurs, the fire is easy to spread along the room, so the horse head wall can effectively arrest the spread of fire (Figure 15-7).

Figure 15-6 Patio

Figure 15-7 Horse head wall

4. North-facing residence

Huizhou general dwellings are mostly north facing south, but most of the dwellings built in the Ming and Qing dynasties in Huizhou are with the front door facing north. The original habits of ancient Huizhou people have many taboos. In this according to the five elements: business is gold, the south is fire, fire grams gold, inauspicious; sign is fire, the north is water, water grams fire, also inauspicious. The merchant in this refers to the merchant, the levy refers to the soldier.

5. Heavy eaves

Heavy eaves, also known as short eaves, is located below the eaves above the doors and windows (Figure 15-8).

6. Full-top bed

"Huizhou full top bed", an all-wooden structure, both Huizhou wood carving, Huizhou lacquerware, Huizhou furniture of the beauty of the old room bed. The so-called "full top" refers to its top, bottom, left wall, right wall and back wall, etc. The five sides are fully sealed with wooden boards and airtight. It is like a small independent house, becoming a "room within a room". In the compartment of Huizhou family, its size is exactly the same as the width of the compartment (Figure 15-9).

7. Pressed painting table

Pressing painting table is the traditional furnishing of Huizhou residence. There are many paintings and couplets hanging on the wall in the middle of the hall of Huizhou houses, which are framed into scrolls and hung. Under the scrolls set up a long table, the desktop placed on two saddle-shaped painting feet, scrolls down to the long table, resting in the painting feet of the "saddle", the painting is flat and stable, the long table called "pressure painting table" (Figure 15-10).

Figure 15-8   Heavy eaves

Figure 15-9   Full-top bed

Figure 15-10   Pressed Painting Table

15.2 Interpretation of the spirit of traditional Chinese culture

The choice of site, layout and architectural form of Huizhou's ancient residential villages are guided by the Zhou Yi theory of feng shui, reflecting the traditional Chinese philosophy of the unity of heaven and man and the aspiration and respect for nature. The elegant Ming and Qing dwelling complexes are closely integrated with nature, creating a living environment that is both scientific and interesting, the essence of traditional Chinese dwellings. The unique water system of the village is a model of water engineering that combines practicality and aesthetics, profoundly reflecting man's remarkable wisdom in using nature and transforming it. It is rare in China for its ancient dwelling complexes to be characterised by their "workmanship, structural ingenuity, beauty of decoration, precision of construction and depth of cultural connotation".

15.3 Project innovation

Review the architectural features of the Huizhou houses in this lesson and make a comparison with the northern houses of Siheyuan studied in the previous course, including the architectural system, style and characteristics, etc., and form a report.

Project 16 Fujian Tulou

16.1 Theoretical guidance

16.1.1 Overview of Fujian Tulou

Fujian Tulou was listed on the World Heritage List in 2008. The "Fujian Tulou" consists of 46 tulou in Yongding, Nanjing and Hua'an counties in Fujian Province. Among them, the 66-meter diameter Jiqing Building has 600 years of "age" and the 31-meter diameter Shanqing Building is only 30 years old.

Fujian Tulou is mainly divided into Yongding Tulou, Nanjing Tulou and Hua'an Tulou, among which the most famous Yongding Tulou is Chengqi Lou, the most famous Tianluokeng Tulou in Nanjing Tulou, and the most famous Hua'an Tulou is Eryilou.

16.1.2 Architectural style and architectural features of Chengqi building

Chengqi building is known as the "King of Round Buildings", with four floors high, four circles, and 400 rooms up and down, with a circle within a circle, and a circle within a circle, which has survived three hundred years of vicissitudes.

The Chengqi Building has a diameter of 73 meters, a corridor perimeter of 229.34 meters, four staircases and 400 rooms. The outer ring building is four floors high, each floor is separated into 72 rooms by a beam-bearing wooden frame inlaid with mud bricks; the second floor does not open windows on the outer wall, but only opens a small window on the inner wall to let in light from the patio; the first floor is the kitchen, and the second floor is the harvest barn, where labor tools and food crops are placed; the third and fourth floors are bedrooms; each floor has a circular passage picked out inwardly, and there are four staircases, which are symmetrically distributed in four directions inside the building. The second ring building has two floors, each with 40 rooms, while the third ring building is a single floor with 32 rooms and the center is the ancestral hall. The third ring building is a single story with 32 rooms, and the center is the ancestral hall. The three ring buildings are like three generals guarding the ancestral hall closely. This is the majesty of the Tulou Grand King, solemn and spectacular (Figure 16–1, Figure 16–2).

Figure 16–1　Exterior of Chengqi Building

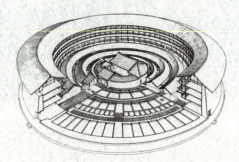

Figure 16–2　Internal structure of Chengqi Building

16.1.3　Building materials and functions of Tulou

Tulou was built to resist floods and wild animals, as well as to defend against foreign invaders.

The special building materials of the Tulou are closely related to its role. The main materials include: clay, cedar, stone, sand, lime, bamboo, tile and other auxiliary materials. By ramming the earth, which means compacting the clay This makes the clay strong, dense and with few gaps, eventually forming a corrosion-resistant, fireproof and earthquake-proof earthen building.

16.1.4　Tianluokeng Tulou cluster

Tianluokeng Tulou is known as "four dishes and one soup", Tianluokeng Tulou Group is mainly composed of one square building, three round buildings and one oval building, respectively, Bu Yun Building, Zhen Chang Building, Rui Yun Building, He Chang Building, Wen Chang Building, square building, Bu Yun Building in the middle and the other four around, these Tulou Group is built in the order of "Gold, Wood, Water, Fire and Earth" five elements (Figure 16–3).

Figure 16–3　Tianluokeng Tulou Cluster

1. Bu Yun Building

Bu Yun Building means that the children and grandchildren will be successful in their studies and will rise to the top of the career ladder (Figure 16–4, Figure 16–5).

Figure 16–4　Buyun Building

Figure 16–5　Interior of Buyun Building

2. Wen Chang Building

Wen Chang Building was established in 1966 and was one of the last earthen buildings to be built. It is the only remaining oval-shaped earthen building, with three floors high and 32 rooms on each floor (Figure 16–6).

3. Zhen Chang Building

Zhen Chang Building was built in 1930, and its distinctive feature is that the inner hall and the door are not in the same line, reflecting the cultural concept of "not revealing the rich". Since there is an ancestral tomb behind the side of Zhen Chang House, the inner hall is set in front of the ancestral tomb, so the direction is staggered from the front door (Figure 16–7, Figure 16–8).

Figure 16–6　Wen Chang Building

Figure 16–7　Zhen Chang Building

Figure 16–8　Interior of Zhen Chang Building

4. Rui Yun Building

Rui Yun Building was built in 1936, it is located in the inner corner of the five buildings, which has the function of hiding wind and gathering qi, reflecting the simple concept of implicit Jishun (Figure 16–9).

5. Hechang building

Hechang building has three floors with 22 rooms on each floor and two staircases (Figure 16–10).

Figure 16–9　Ruiyun building

Figure 16–10　Hechang building

16.1.5　Eryi Building

The building was built 30 years ago and covers an area of 9,300 square meters, with a diameter of 73.4 meters, and is the largest double-ring circular earth building of its kind in Fujian Province. The inner ring bungalow is a transparent house with kitchen and dining room. Each unit of the outer ring building has four rooms each. The ground floor is bedroom and living room; the second and third floors are bedrooms and warehouses, and the fourth floor is the ancestral hall of each unit.

The structure of the outer wall of Eryi Building is the same as that of Chengqi Building of Yongding Tulou before, and the thickness of the outer wall is very thick (Figure 16–11).

Figure 16–11　Internal structure of Eryi Building

The special feature of Eryi Building is the fire prevention system of the gate, which can be extinguished by injecting water through bamboo tubes if the gate is attacked by fire, which fully reflects the wisdom of Hakka people(Figure 16–12).

In the center of Eryi Building is the inner courtyard of the public place, where clothes and crops can be dried on the grounds, and there are two ancient wells in the courtyard, named Yin Quan and Yang Quan, forming the formation of Tai Ji(Figure 16–13).

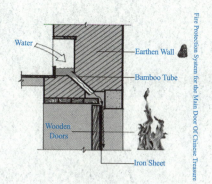
Figure 16-12　Fire protection syetem of Eryi Gate

Figure 16-13　Yin Quan & Yang Quan

16.2　Interpretation of the spirit of traditional Chinese culture

The unity of heaven and man in the architectural concept of Tulou, the livability of the building form, and the harmony and spirituality of the cultural meaning highlight the unique style of traditional Chinese residential architecture and the wisdom of the Hakka people, and its ecological, environmental protection, and open architectural ideas have greater significance for the creation of modern architecture.

16.3　Project innovation

Review the contents of this section of the course, compare the similarities and differences between Chengqi Building and Eryi Building for example, you can compare them in terms of function, construction, building materials, layout features, etc., and form a PPT report assignment.

Project 17 Architectural elements in the garden

17.1 Theoretical guidance

17.1.1 Architectural elements and characteristics in classical Chinese gardens

The architectural elements in Chinese gardens include: Pavilions, Terrace, Storeyed Building, Tower, Windowed veranda, waterside pavilion, corridor and boat. These words are difficult to find corresponding words in English, and are generally called pavilions or buildings, because they think that the architectural elements in these gardens are almost similar, and their functions are not the same. In fact, in ancient architecture, there are clear definitions and distinctions between them, and their functions vary according to their location, size, and form, as well as their ornamental aesthetics.

1. Pavilions

The pavilion, also known as a gazebo, is a traditional Chinese building that originated from the Zhou Dynasty. Most of them are built on the roadside for pedestrians to rest, take a cool break or enjoy the view. The pavilion is generally an open structure without walls, and the building form is characterized by a roof without walls. The top can be divided into hexagonal, octagonal, round and other shapes. They are widely used in garden architecture because of their light shape, unconstrained choice of materials and flexible layout (Figure 17-1).

2. Terrace

Terrace originated in the Shang Dynasty and flourished in the Spring and Autumn and Warring States Periods, and is the oldest form of garden architecture, with many ancient palace buildings built on top of it. The early stage is a kind of towering rammed earth building for overlooking, its function is mainly for sacrifice, but also has the function of viewing and entertainment. Later, it evolved into an open-air platform in front of the hall, namely the moon platform. There is also a theater platform, which was mainly used for singing and entertainment in ancient times (Figure 17-2).

3. Storeyed Building

The Storeyed building is a house of two or more floors in ancient architecture, also called a Overlapping building. Building has appeared in the Warring States period, the Han Dynasty city building has been as high as three floors (Figure 17-3).

Figure 17-1   Pavilions

Figure 17-2   Terrace

Figure 17-3   Building

4. Tower

We often say that the Overlapping house is the building, four open for the pavilion. All four sides of the pavilion have windows, and also has a door, surrounded by a flat seat, for people to walk around the pavilion, the view also has a collection of books and for Buddhist use. We often talk about the "pavilion", the pavilion and the building always seem to appear together. Compared with the building, the pavilion is smaller in size (Figure 17-4).

5. Windowed veranda

Windowed veranda refers to a small and bright building in the garden, suitable for reading. It is also called "fast" or "room". The ancient meaning of Windowed veranda is a promenade or hut with windows. Mostly high and open buildings, but the volume is not large. Windowed veranda form type also more, different shapes, as wide as the corridor, is a kind of embellishment of the building (Figure 17-5).

6. Waterside pavilion

The waterside pavilion is built on a high earthen platform or water or the building on the water, mostly refers to the water pavilion, where there is a pavilion will be water, pavilion is for people to rest, viewing the building. More bar and sitting stool railings, these railings and sitting stool is also known as beauty leaning, Wu Wang leaning. According to legend, it is the King of Wu and the beauty of the King of Wu and Xi Shi travel to enjoy the scenery of things (Figure 17-6).

Figure 17-4　Windowed veranda　　　Figure 17-5　Windowed veranda　　　Figure 17-6　waterside pavilion

7. Corridor

The corridor is a passage connecting two buildings. There is a roof on it, supported by columns, to shade the sun and keep out the rain, and to facilitate people to view the scenery during their wandering. In terms of architectural art, it increases the spatial level (Figure 17-7).

8. Boat

Boat is a boat-shaped building built on the water surface of the garden after the shape of a boat, and boat is the meaning of boat. It looks like a boat but cannot be paddled, so it is also called "untethered boat". Most of them face the water on three sides and are connected to the land on one side for people to play and set up banquets and enjoy water scenery (Figure 17-8).

Figure 17-7　Corridor　　　　　　　　Figure 17-8　Boat

17.1.2　Four Famous Pavilions in China

There are four famous pavilions in China: Drunken Master Pavilion, Tao Ran Pavilion, Aiwan Pavilion, Hu Xin Pavilion.

1. Drunken Master Pavilion

Drunken Master Pavilion, the first of the four famous pavilions in China, is also known as the first pavilion in the world. It is located in the Langya Mountain Scenic Spot in the southwest of Chuzhou City, Anhui Province (Figure 17-9).

2. Tao Ran Pavilion

Tao Ran Pavilion is located in Beijing, when Jiang Zao, the then Minister of Public Works, was ordered to supervise the black kiln factory, he built a small pavilion in the western part of Cibeian, and took the word "Tao Ran" in the line of Bai Juyi's poem "Waiting for the chrysanthemum yellow home brew to be ripe, and a drunken Tao Ran with you" to name the pavilion (Figure 17-10).

Figure 17-9　Drunken Master Pavilion　　　　Figure 17-10　Tao Ran Pavilion

3. Aiwan Pavilion

Aiwan Pavilion is located in Changsha, Hunan Province, originally named "Red Leaf Pavilion", also known as "Love Maple Pavilion". The name was changed to Ai Evening Pavilion according to the poet Du Mu's "Mountain Journey" in the Tang Dynasty, which is a poetic meaning of "Sitting in the maple forest at night, the frost leaves are redder than the flowers in February" (Figure 17-11).

The architectural style of the pavilion is eight columns with heavy eaves, covered with green glazed tiles on top, with a pointed pavilion roof, red wooden columns on the inner columns, granite square columns on the outer columns, and painted algae wells on the interior.

4. Hu Xin Pavilion

Huxin Pavilion is located in the center of West Lake in Hangzhou, Zhejiang Province, and is called "Huxin Pingtiao" in the 18 scenic spots of West Lake (Figure 17-12).

Figure 17-11　Aiwan Pavilion　　　　Figure 17-12　Hu Xin Pavilion

17.1.3　Three Great Buildings of Southern China

The three most famous buildings in South China include Yellow Crane Tower in Hubei, Yueyang Tower in Hunan and Tengwang Pavilion in Jiangxi.

1. Yellow Crane Tower

Yellow Crane Tower is located in Wuhan City, Hubei Province, and has now become a landmark building in Wuhan City, as one of the "three famous buildings in South China". Yellow Crane Tower main building for four sets of octagonal body, 51.4 meters in height, the bottom side of 30 meters wide, the top side of 18 meters wide, five eaves, save the tip of the roof, covered with golden glazed tiles, supported by 72 columns, upstairs there are 60 angles to the outside.

The appearance of Yellow Crane Tower has five floors, but the interior actually has nine floors, which implies the meaning of nine to five, and the crane wing shape with eight flying eaves reflects the unique culture of Yellow Crane Tower, making the perfect combination of Chinese traditional architectural features and cultural meaning (Figure 17–13).

2. Yueyang Tower

The main building of Yueyang Tower is rectangular in shape, 19.42 meters high, 14.54 meters deep and 17.42 meters wide, with three floors, four columns, flying eaves, helmet roof and pure wooden structure. On the second floor of the building, there is a carved screen with poems, couplets and carvings; on the second floor, there is a carved screen of paulownia wood, on which is engraved the "Record of Yueyang Tower" written by Zhang Zhaoshu, a calligrapher of the Qing Dynasty; on the third floor, there is a carved screen with the poem "Climbing Yueyang Tower" written by Du Fu in Mao Zedong's handwriting, and a couplet of "Changgeng Li Bai" is hung on the eaves pillar "The water and the sky are one color, the wind and the moon are boundless", which has certain ornamental value. As the only ancient building among the three famous towers, the unique structure of the helmet roof reflects the wisdom of ancient working people and the skillful design of skilled craftsmen (Figure 17–14).

3. Tengwang Pavilion

Tengwang Pavilion, located in Nanchang City, Jiangxi Province, is a landmark building in Nanchang City. The Tengwang Pavilion was built by Li Yuanying, the brother of Emperor Taizong Li Shimin, when he was the governor of Hongzhou in Jiangnan, and was known as the "First Building on the West River" (Figure 17–15).

Figure 17–13 Yellow Crane Tower

Figure 17–14 Yueyang Tower

Figure 17–15 Tengwang Pavilion

The main building of Tengwang Pavilion is 57.5 meters high, with a construction area of 13,000 square meters; the lower part of the building is a 12-meter-high pedestal symbolizing the ancient city wall, divided into two levels; the main pavilion above the pedestal is in the format of "bright three dark seven", with a three-story corridor in imitation of Song-style architecture, and a total of seven floors inside, divided into three bright floors, three dark floors and the attic.

17.2 Interpretation of the spirit of traditional Chinese culture

The concept of "unity of heaven and man" is the soul of classical Chinese garden art. Its main point is the pursuit of respect for nature under the premise of transformation of nature, to create a pleasant and harmonious garden ecology. In today's world, where industrial civilization threatens the living environment of human beings, this ideology itself contains some kind of solution. This ideology itself contains some kind of solution and philosophical reasoning, which is worthy of our learning and reference, especially in modern planning and design can inherit this great idea.

17.3 Project innovation

Review the content of this lesson, review and organize the knowledge of "Pavilions, Terrace, Storeyed Building, Tower, Windowed veranda, waterside pavilion, corridor and boat" and present them in the form of a mind map.

Project 18 Elements of the garden

18.1 Theoretical guidance

The landscape elements in the garden mainly include: mountains, water, plants, animals, buildings, couples, plaques, stone carvings, which are the eight major elements. According to the traditional Chinese art law for design and combination, so as to reflect the natural spiritual realm of garden art as the ultimate and highest purpose, so as to achieve the "although made by man, just from heaven" aesthetic interest.

1. Stacked mountains

Stacked mountain is to take the true mountain of the mountain posture and appearance, momentum and rhythm, after the art of generalization, refinement, and now in the garden, the shape of a small mountain to pass the god of the mountain. This kind of mountain created by some real mountain mood, will give people a sense of intimacy, with a rich imagination and taste (Figure 18-1).

Traditional stone selection criteria for stacked mountains are transparent, leaky, thin, wrinkled and ugly. The so-called "transparent, that is, exquisite porous, the shape of the contour of the multi-colored"; thin, thin in the show; wrinkled, the shape of the ups and downs, light and dark more change, rich in a sense of rhythm; leak is the peak of the stone up and down around the trick, there is a way to follow. The ugly stone peak is quite imposing, can create a strong and ancient mood, to attract people's imagination.

The texture of the stone can be divided into the following categories:

Lake stone, originally produced in Taihu Lake area and named, is the most commonly used in the garden, but also the history of the early development of a class of mountain stone. Stone surface porous, geological hard and brittle, texture vertical and horizontal. Its subcategories are Taihu stone, Fangshan stone, Yingde stone, Yellow stone, Stalagwites, etc, Lingbi stone, Xuan stone. Yellow stone, a yellowish fine sandstone, has more origins. It is angular, majestic and solid. Stalagmites, with the shape of long bamboo shoots, are widely produced, lying in the mountains and standing upright on the ground after mining (Figure 18-2).

Taihu Lake stone and yellow stone is the most common stone.

Figure 18-1 Stacked mountains

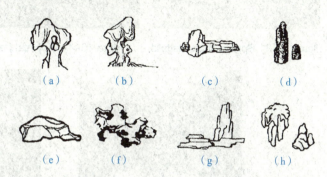

Figure 18-2 Stone selection criteria
(a) Taihu Stone; (b) Lingbi Stone; (c) Yellowstone;
(d) Stalagmite; (e) Xanthophyllite; (f) Yingde Stone;
(g) Xuanshi; (h) Stalactite

There are three major functions of stacked hills in garden art. Firstly, it can organize and divide and separate the space, secondly, it can coordinate the environment according to the local conditions, and at the same time, it can also embellish the landscape.

2. Water management

Water management refers to all kinds of water treatment in the garden. Water is flowing, indeterminate, and the stability of the mountain, fixed to form a sharp contrast. Traditional garden water is a generalization, refinement and reproduction of the characteristics of natural landscape. The performance of the various types of water form, not in the absolute volume close to nature, but in the artistic reality of the landscape features.

Garden water management has the following common types: springs and waterfalls, abyssal ponds, streams, rivers, ponds, lakes.

Springs for the underground gushing water, waterfall is a cliff fall of water. The water source is either a natural spring, or outside the park to draw water or artificial water. Waterfall has a linear, curtain, diversion, stacked fall and other forms; abyssal pond is a small and deep body of water, dark light, low water level; stream is a dynamic waterscape of water flowing from the mountains of springs and waterfalls, with gravel as the bottom, the stream is shallow, you can count the swimming fish, and can wade;. Rivers in the garden is often used in narrow pools to show, so that the scenery is rich in change. Rivers can be long or short, straight or curved, wide or narrow, there is a collection and release. Rivers are mostly used with earthen banks, configured with appropriate plants; lakes are patches of converging water (Figure 18-3— Figure 18-7).

Figure 18-3　Springs and waterfalls　　Figure 18-4　Abyssal ponds　　Figure 18-5　Streams

Figure 18-6　Abyssal ponds　　Figure 18-7　Abyssal ponds

The role of water bodies in gardens can not only create scenery, but also drainage and water storage, improve the ecological environment, regulate the microclimate of the local environment and carry out various activities and tours on water.

3. Plant elements

Plants are an indispensable element of the water, water features are not beautiful if you leave the flowers and trees, with these flowers and trees to make the mountains greener and the water more beautiful.

First of all, we should pay attention to plant diversity. Nature's plants are varied and colorful, and have a good ornamental value in themselves. Secondly, the layout is reasonable, sparse and orderly, combining single groups. Natural plants are not all grouped, there are also solitary. Garden plant configuration on the lone planting, planting, piece planting, group planting, mixed planting in a variety of ways, so that not only to appreciate the poise of the lone tree, but also to appreciate the beauty of the group planting trees. In addition to pay attention to the reasonable matching of different garden plant forms and colors, the configuration of garden plants should be planted according to the topography of different forms and colors of plants, and between them can not cause a conflict of perspective. Finally, we should pay attention to the cultural integration of garden plants with the surrounding environment. For example, the combination with the road and the combination with the architecture (Figure 18-8, Figure 18-9).

4. Animal elements

Animals in gardens are mainly for people to enjoy, to better show the integration with the natural environment and to produce a sense of spiritual atmosphere. The animals in the garden mainly include: birds, ducks, fish, etc., where fish occupy the primary position among the garden animal elements (Figure 18-10).

Figure 18-8　Garden plants（1）

Figure 18-9　Garden plants（2）

Figure 18-10　Fish in the garden

5. Architectural elements

The architectural elements in Chinese gardens include: Pavilions, Terrace, Storeyed Building, Tower, Windowed veranda, waterside pavilion, corridor and boat.

6. Plaques and Couplets

Plaques and couplets are mainly hung above halls, halls, pavilions and pavilions. The plaque is placed across the door or on the door of the wall, and is mostly the name of the attraction or praise for the scenery in the garden; the couplets are often matched with the plaque and hung on both sides.

Plaque couplet can not only embellish the hall and pavilion, decorate the door wall, in the garden often express the thoughts and feelings of the garden maker or garden owner, but also can enrich the landscape, evoke the association, increase the poetic mood, play a role in the finishing touch, is a feature of the traditional Chinese garden (Figure 18-11, Figure 18-12).

Figure 18-11　Plaques and Couplets（1）

Figure 18-12　Plaques and Couplets（2）

7. Stone carving

The carved stone is characterized by its beauty, roundness, thinness, thickness, uprightness and beauty. It is a kind of landscape stone and garden stone, and is an important carrier for engraving poems and inscriptions. The organic combination of the natural rustic shape and texture of the stone and the dynamic and painterly lines of Chinese calligraphy is itself a symbol of the unity of human and nature (Figure 18–13, Figure 18–14).

Figure 18–13　Stone carving（1）　　　Figure 18–14　Stone carving（2）

18.2　Interpretation of the spirit of traditional Chinese culture

The mountains, water, plants, animals, architecture, couples, plaques and stone carvings that make up the elements of Chinese gardens are the stuff of legend in Chinese architecture. It is also a wonder of the world of art and an important heritage of human civilisation. It embodies the spirit of Chinese national culture, harmonising the idea of nature with the spirit of humanity. It is the crystallisation of the wisdom and creativity of the ancient Chinese working people, and the comprehensive embodiment of ancient Chinese philosophical thought, culture and art.

18.3　Project innovation

Review the content of this lesson, talk about one of the component elements that interests you most, and complete the PPT assignment.

Project 19　Royal Gardens

19.1　Theoretical guidance

19.1.1　Overview of the Royal Gardens

The Summer Palace, a royal garden during the Qing Dynasty in China, formerly known as the Qing Ripple Garden, is located in the western suburbs of Beijing and covers an area of 3.009 square kilometers, with about three-quarters of the water surface.

19.1.2　Functions and main architectural layout of the Summer Palace

As an important part of the royal living environment, the Summer Palace was a place for the emperor to take a rest or deal with government affairs. The Summer Palace is mainly composed of two parts: Wanshou Mountain and Kunming Lake. There are more than 3,000 palace and garden buildings in various forms, which can be roughly divided into three parts: administration, living, and sightseeing.

19.1.3　The main architectural elements in the Summer Palace and their functional roles

1. Suzhou Street

Suzhou Street, also known as the "Street of Sale and Purchase", was built to imitate the water town in Jiangnan. During the Qing Yi Yuan period, there were various stores on the shore, such as jade and antique stores, silk stores, dim sum stores, teahouses, gold and silver jewelry buildings. Shopkeepers in the stores are eunuchs, palace maid makeup. The emperor started "business" when he visited. Qianlong emperor's first tour of Jiangnan, because of the lively streets and stores and products and customs of Suzhou, Jiangnan, so the accompanying painters in accordance with this form of drawing style, but also the marketplace culture into the royal architecture, which the marketplace refers to the common people (Figure 19–1).

2. Wanshou Mountain

Wanshou Mountain is the remnant of Yanshan Mountain, 58.59 meters high and 108.94 meters above sea level. From the "Yunhui Yu Yu" pagoda at the foot of the mountain, through Paiyun Gate, Er Gong Gate, Paiyun Hall, Dehui Hall, Buddha Fragrance Pavilion, to the Wisdom Sea at the top of the mountain, it forms a central axis that rises in layers (Figure 19–2).

Figure 19–1 Suzhou Street

Figure 19–2 Wanshou Mountain

3. The Four Continents

The Four Continents of the Summer Palace was built according to ancient Indian philosophical concepts. It is a group of Chinese and Tibetan buildings in the middle of the back hill of Wanshou Mountain in the Summer Palace, covering an area of 20,000 square meters. The imperial family during the Qing Dynasty all believed in Buddhism especially during the late Qing Dynasty, hence the construction of this place.

The Four Great Continents refers to the four Buddhist continents, which are interpreted in Sanskrit and include the North Jurassic, South Nanzhan, East Shenzhou, and West Niugata continents, which are square, triangular, half-moon, and circular in shape, and correspond to the "four great" Buddhists, namely, earth, fire, wind, and water (Figure 19–3).

The main buildings in the four continents are sutra pillars and lama pagodas. The sutra pillar is a Buddhist decoration with many small statues of Buddha on it and also engraved with Vajra Sutra, which is 3 meters high; in the south, southwest, northeast and northwest of the four continents, there are four lamashtas representing the "four wisdoms" of Buddhist scriptures: red, white, black and green (Figure 19–4—Figure 19–8).

The black pagoda is "equal wisdom", which means to see all things in the world as equal and undifferentiated; the white pagoda is "great circle wisdom", which means to clearly reflect the understanding of all things in the world; the red pagoda is "wonderful observation wisdom". The red pagoda is "Myriad Observation Wisdom", which means that the Buddha is able to perceive good and evil, and to observe all dharmas; the green pagoda is "Becoming Wisdom", which means that believing in the Buddha can achieve self-benefit and benefit others.

Figure 19-3　The Four Continents　　　　　　Figure 19-4　Sutra Banner

Figure19-5　Black pagoda　　Figure19-6　White pagoda　　Figure19-7　Red pagoda　　Figure 19-8　Green pagoda

4. Buddha Fragrance Pavilion

Buddha Fragrance Pavilion is an eight-sided, three-story, four-eaves building; the pavilion is 41 meters high, with eight huge iron pear wood pillars inside the pavilion, the structure is complex and is a classical architectural masterpiece. The pavilion is used for the royal family to perform pujas here (Figure 19-9).

5. Baoyun Pavilion

Bao Yun Pavilion is commonly known as the Copper Hall or Copper Pavilion. Shape after the style of wooden buildings, heavy eaves hillside roof. Bao Yun Pavilion 7.55 meters in height, weighing about 207 tons. Hall components pillars, beams, rafters, tiles, ridge-kissed beasts, even plaques, etc. are like wooden structures. It sits on a carved Chinese white jade Sumeru seat (Figure 19-10).

6. Paiyun Temple

Paiyun Temple was originally built by Qianlong for his mother's 60th birthday and the Grand Retribution Temple, and was renamed Paiyun Hall when Cixi rebuilt it. The word "Paiyun" is taken from Guo Pu's "Poem on Traveling Immortals" of the Jin Dynasty: "The gods came out of the clouds, but I saw a gold and silver platform" (Figure 19-11).

Appreciation of Chinese Traditional Architecture

Figure 19-9　Buddha Fragrance Pavilion

Figure 19-10　Baoyun Pavilion

7. Promenade

Promenade is located in the southern foot of Wanshou Mountain 728 meters long, is the longest corridor in the Chinese garden, the promenade is also a colorful gallery, the corridor between each square beam are painted with colorful paintings, a total of more than 14,000, bright colors, rich and beautiful. Color painting Chinese wooden architecture on the decorative arts. It is characterized by: the main picture is bracketed in a large semicircle of brackets（called "package"）; no fixed structure, all depending on the painter to play, the same subject matter can create different images. Promenade paintings have a wide range of subjects, including mountains and forests, flowers and birds, scenery and people. Among them, the most fascinating is the story paintings of people, with more than 2,000 paintings in total, no two of which are the same, and the painters have concentrated thousands of years of Chinese history and culture on this long corridor (Figure 19-12—Figure 19-15).

Figure 19-11　Paiyun Temple

Figure 19-12　Promenade

Figure 19-13　Soviet-style color painting（1）

Figure 19-14　Soviet-style color painting（2）

Figure 19-15　Soviet-style color painting（3）

8. Grand Opera House

Grand Opera House in the Summer Palace is the largest surviving ancient opera house in China. It is 21 meters high and divided into three floors: upper, middle and lower. The center of the lower ceiling is connected to the upper stage by a patio, and the middle stage is equipped with a winch, which can be set up by a tricky mechanism, so that it can go up to the sky and enter the ground with endless changes. The construction of the Grand Opera House is because Empress Dowager Cixi was fond of Peking Opera (Figure 19–16).

9. Qingyan Boat

Qingyan Boat is a big stone boat, also called stone boat, with a total length of 36 meters. The boat is carved and stacked with marble, Qing Yan Boat is a famous architectural treasure on the water in the Summer Palace, and is the only building with Western style in the Summer Palace. The boat is built with a two-story boat building, with flower tiles on the bottom of the boat, stained glass windows and brick decoration on the top (Figure 19–17).

Figure 19–16 Grand Opera House   Figure 19–17 Qingyan Boat

10. Bronze Bull

Bronze Bull is an architectural sculpture located on the north side of the east bridge of the 17-hole bridge, which was cast in bronze in 1755, the 20th year of the Qianlong era, and is called the "Golden Bull" (Figure 19–18).

11. Seventeen Holes Bridge

Seventeen Holes Bridge is a masterpiece of ancient bridge architecture. The bridge consists of 17 holes, 150 meters long, with 544 lions of different sizes carved on the pillars of each bridge railing. The whole 17-hole bridge has great aesthetic value, academic value and usage value (Figure 19–19).

Figure 19–18 Bronze Bull   Figure 19–19 Seventeen Holes Bridge

19.2 Interpretation of the spirit of traditional Chinese culture

The Summer Palace is a masterpiece of traditional gardening art, borrowing from the surrounding landscape environment, with the grandeur of royal gardens, but also full of natural interest, highly embodying the Chinese garden "although made by man, just like from heaven" guidelines for gardening. In December 1998, the Summer Palace was inscribed on UNESCO's World Heritage List and was evaluated as follows: The Summer Palace in Beijing is an outstanding demonstration of the art of Chinese landscape gardening, which harmoniously blends man-made landscapes with nature; the Summer Palace is a concentrated expression of Chinese gardening thought and practice, which has played a key role in the development of the entire Eastern garden art and culture. The Summer Palace is a powerful symbol of one of the world's great civilizations, represented by the Chinese royal gardens.

19.3 Project innovation

Review the contents of this lesson and write a reflection on the Summer Palace, the royal garden.

Project 20　Private Garden

20.1　Theoretical guidance

20.1.1　Construction characteristics of private garden

The first characteristic of private garden is the small scale, generally only a few acres to a dozen acres, the smallest only one and a half acres. The main idea of the gardeners is to see the big in the small, that is, within a limited scope of the use of subtlety, depression, twists and turns, hints and other techniques to start the subjective re-creation of people, twists and turns, resulting in a seemingly deep and endless scenery, expanding people's feelings about the actual space; secondly, the construction of the water surface, private gardens are mostly centered on the water surface, surrounded by scattered buildings, constituting a spot, several spots surrounded by a scenic area. The third point is to cultivate the body, the main function of the garden is to cultivate the body, leisure and self-entertainment for; the fourth point is the nobility and elegance, the garden owner is mostly literati bachelor origin, rather than builders or professionals to build, literati bachelor can poetry and painting, good at tasting, garden style to the nobility and elegance, light vegetation off-beat for the highest pursuit, overflowing with a rich sense of books and scrolls; the fifth point is the gardening techniques in the layout of the use of flexible and versatile techniques, good at The fifth point is that the gardening techniques are flexible in layout, good at imitating the image of natural landscapes, and pay much attention to the detailed treatment of gardens.

20.1.2　The Four Famous Gardens of China

The Humble Administrator's Garden, together with the Summer Palace in Beijing, the Summer Palace in Chengde, and the Liuyuan Garden in Suzhou, is known as one of the four most famous gardens in China.

20.1.3　Scenic techniques in gardens

1. Borrowed scenery

Borrowed scenery is one of the common framing techniques used in classical garden architecture. It is a technique of organizing a good view into the garden sight within the reach of the eyes. Architecture, water, rocks, plants and animals are interspersed with each other (Figure 20-1).

2. Adding scenery

Adding scenery refers to the space is more empty, the landscape is more monotonous, the lack of depth of field, add a scene to change the situation of the composition of the technique, mainly to play a transitional role. It is equivalent to adding a middle scene between the distant scene and the near scene (Figure 20-2).

3. Framing scenery

Framing scenery on the limited space to harvest the infinite space of the partial picture of the composition of the scenery. Such as ancient gardens, commonly used doors, windows, holes, pavilions, columns, eaves, bars, etc. constitute the frame boundary composition, so as to highlight the local landscape or the effect of moving the scene (Figure 20-3).

Figure 20-1　borrowed scenery　　Figure 20-2　Adding scenery　　Figure 20-3　Framing scenery

4. Leaky scenery

Leaky scenery is developed from frame view. The framed view is all-viewing, while the leaking view is hidden and elegant. Leakage scenery can be used with leaky windows, leaky walls, leaky screens, sparse forests and other techniques (Figure 20-4).

5. Opposite scenery

Opposite scenery refers to a certain space between two scenes, so that the viewer can view the scenery of the other place from here.For example, the two pavilions between the remote reflection (Figure 20-5).

6. Suppression of scenery

Suppression of scenery types are mountain suppression, tree suppression, curved suppression. Chinese traditional art has always been subtle, avoid a glance, a look to the edge.

So landscape gardening will never let people walk into the door to see the best view, the best scenery is often hidden behind, which is called "first hidden after revealing", "want to raise first to suppress". Such as the entrance to the garden often welcome the door to block the rockery, this treatment is called "mountain suppression", the entrance has a tree to block, called tree suppression (Figure 20–6).

Figure 20–4　Leaky scenery

Figure 20–5　Opposite scenery

Figure 20–6　Suppression of scenery

20.1.4　Plants in the garden

1. Bamboo

The language of bamboo is likened to that of a gentleman with a temperament, symbolizing steadfastness, high-mindedness and modesty (Figure 20–7).

2. Plum blossom

Plum blossom symbolizes noble, strong and modest character, and give people an inspirational spirit of striving to be strong (Figure 20–8).

3. Apricot tree

Ginkgo symbolizes longevity, fortitude and composure, and also represents eternal love (Figure 20–9).

Figure 20–7　Bamboo

Figure 20–8　Plum blossom

Figure 20–9　Apricot tree

4. Pine tree

The pine tree is an evergreen tree with strong vitality, symbolizing strong will and unyielding character, and also a symbol of longevity (Figure 20–10).

5. Chrysanthemum

The chrysanthemum blooms in the frost and is proud of its bones, symbolizing noble and strong sentiments (Figure 20-11).

Figure 20-10 Pine tree   Figure 20-11 Chrysanthemum

6. Lotus

The lotus flower, "out of the mud but not dyed, wash the pure lotus but not demon, the middle pass outside straight", the lotus flower metaphor for the gentleman, symbol of holiness (Figure 20-12).

7. Pomegranate tree

The pomegranate tree is a symbol of many children and grandchildren (Figure 20-13).

Figure 20-12 Lotus   Figure 20-13 Pomegranate tree

20.2 Interpretation of the spirit of traditional Chinese culture

Classical Chinese gardens have been created by generations of craftsmen to build natural and beautiful scenery with artificial design to embellish the environment, forming a unique style of traditional cultural and artistic features, borrowing scenery, adding scenery, framing scenery, leaking scenery, pairing scenery, suppressing scenery gardening techniques can be described as ingenious, to the literati and elegant people to build the art of gardening in the history of Chinese ancient architecture has left a colorful stroke, humanistic spirit and the integration of humanistic sentiment also makes Garden art is like a wisp of green ink emitting a unique fragrance. The symbolic meaning of each plant is integrated with the mood of the garden, making the garden and the natural world as if they were born.

20.3 Project innovation

Review the contents of this lesson and talk about the style of literati gardening as reflected in the gardens of Suzhou.

Project 21　Zhongnanhai

Theoretical guidance

1. Overview of Zhongnanhai Architectural

Zhongnanhai, formerly known as Xiyuan and Taiyanchi, is located on the north side of West Chang'an Street in the Xicheng District of Beijing, China, west of the Forbidden City and Jingshan Mountain, and is the collective name for Zhonghai and Nanhai, which together with Beihai are known as the Three Seas.

Zhongnanhai covers a total area of about 1 million square meters, of which the water area is about 460,000 square meters. It is mainly composed of buildings and gardens such as Ziguangge, Qinzheng Hall, Yiluan Hall, Jiaoyuan Garden, Shui Yunshe, Yingtai, Fengze Garden, Jinggu, etc. It has both solemn palace-style architecture and flexible garden-style architecture, and is one of the more complete royal palaces in China (Figure 21-1).

Zhongnanhai was built in the Liao Dynasty and became a royal palace during the Ming and Qing Dynasties. It is now the seat of the State Council of the People's Republic of China, the Secretariat of the Central Committee of the Communist Party of China and the General Office of the Central Committee of the Communist Party of China, and is one of the landmarks and national symbols of China.

Yingtai is an immortal island palace located in the South Sea of Zhongnanhai. It was first built in the Ming Dynasty, and then twice during the Shunzhi and Kangxi Dynasties of the Qing Dynasty, as a place for emperors and consorts to listen to government, take refuge and live. It is named Yingtai because it faces the water on all sides and is lined with pavilions and pavilions like a fairy island in the sea (Figure 21-2).

The Hundred Days' Reform was a patriotic movement with the meaning of saving the country, an important political reform in modern Chinese history, and an ideological enlightenment movement, which promoted the emancipation of the mind and played an important role in the development of thought and culture and the progress of modern Chinese society.

Figure 21-1   Zhongnanhai

Figure 21-2   Yingtai in Zhongnanhai

2. Establishment of New China

With the failure of the Hundred Days Reform and the change of history, the Republic of China was established in 1912, and after the Qing Emperor's abdication, Zhongnanhai was occupied by Yuan Shikai and became the presidential palace of the Beiyang government (Figure 21-3).

Zhongnanhai was used as the office of the president and premier of the Beiyang government, as well as Zhang Zuolin's "handsome residence" (Figure 21-4). After the Nationalist government moved its capital to Nanjing, Zhongnanhai was opened to the public as a park.

After 1949, Zhongnanhai became the residence of the Central Committee of the Communist Party of China and the State Council of the People's Republic of China, as well as the residence of some national leaders.

From September 21 to 30, 1949, a plenary session of the Chinese People's Political Consultative Conference was held in the magnificent Huairen Hall in Zhongnanhai. Huairen Hall was originally a huge courtyard in the imperial court of the Qing Dynasty, with a central patio. The architect Liang Sicheng, on the order of Mao Zedong, built an auditorium on the patio that could accommodate a thousand people, which was seamlessly integrated with the original palace's ancient architecture, saving time and money and having national characteristics, showing the artist's ingenuity, and became the hall where Mao Zedong and the CPPCC delegates discussed the grand plan for the founding of the country (Figure 21-5).

Figure 21-3   Xinhua Gate

Figure 21-4 Zhang Zuolin's "Office of the Marshal"

Figure 21-5 Huairen Hall

At 2 p.m. on October 1, 1949, Mao Zedong presided over the first meeting of the Central People's Government Committee in Zhongnanhai, and the Chairman, Vice Chairman and members of the Central People's Government were inaugurated, and the Central People's Government was established.

参考文献

［1］田永复.中国古建筑知识手册［M］.2版.北京：中国建筑工业出版社，2019.
［2］刘敦桢.中国住宅概说·传统民居［M］.武汉：华中科技大学出版社，2018.
［3］楼庆西.中国古建筑二十讲［M］.北京：生活·读书·新知三联书店，2001.
［4］张克群：中国古建筑小讲［M］.北京：化学工业出版社，2020.
［5］杜爽.中国古建筑工程技术系列丛书——古建筑油漆彩画［M］.北京：化学工业出版社，2012.
［6］宫殿建筑（末代皇都）［M］.北京：中国建筑工业出版社，2010.

References

[1] Tian Yongfu. *Chinese Ancient Architecture Knowledge Manual* [M].2 edition. Beijing: China Architecture and Building Press, 2019.

[2] Liu Dunzhen. *Introduction to Chinese Housing: Traditional Dwellings* [M]. Wuhan: Huazhong University of Science and Technology Press, 2016.

[3] Lou Qingxi. *Twenty Lectures on Ancient Chinese Architecture* [M]. Beijing: Life, Reading, new Knowledge Sanlian Bookstore, 2001.

[4] Zhang Kequn. *Lecture on Ancient Chinese Architecture* [M]. Beijing: Chemical Industry Press, 2020.

[5] Du Shuang. *Chinese Ancient Architecture Engineering Technology series—Ancient Architecture Paint Painting* [M]. Beijing: Chemical Industry Press, 2012.

[6] China Architecture and Building Press. *Palace Architecture (Capital of the Last Emperor)*. Beijing: China Architecture and Building Press, 2010.